AN INTRODUCTION TO DATA ANALYSIS

An Introduction to Data Analysis

Bruce D. Bowen
BLUE CROSS OF NORTHERN CALIFORNIA

Herbert F. Weisberg
THE OHIO STATE UNIVERSITY

W. H. Freeman and Company
San Francisco

Sponsoring Editor: Richard J. Lamb
Project Editor: Larry Olsen
Production Coordinator: Fran Mitchell
Compositor: Typothetae
Printer and Binder: The Maple-Vail Book Manufacturing Group

Library of Congress Cataloging in Publication Data

Bowen, Bruce D
An introduction to data analysis.

Bibliography: p.
Includes index.
1. Mathematical statistics. I. Weisberg, Herbert F., joint author. II. Title.
QA276.B674 001.4'225 79-27870
ISBN 0-7167-1173-7
ISBN 0-7167-1174-5 pbk.

9 8 7 6 5 4 3 2 1

Contents

Preface ix

Reference Tables xi

1 **The Role of Data Analysis** 1

The Importance of Data Analysis 2

What Is Data Analysis? 5

2 **A Review of Research Design** 6

Three Basic Problems of Scientific Research 6

Three Types of Studies 8

The Quality of the Measurements 11

Sampling Procedures 12

Summary 17

Questions 17

3 **The Analysis Process** 18

Explanation 18

Causal Processes 22
Planning an Analysis 24
Summary 27
Questions 27

4 **Computer Utilization** 28
Data Storage 28
Data Analysis 31
Summary 37
Questions 37

5 **Single-Variable Statistics** 38
Frequency Distributions 38
Levels of Measurement 46
Measures of the Central Tendency 49
Measures of Dispersion 53
Statistical Inference for Means 55
Summary 57
Questions 57

6 **Two-Variable Tables** 58
Reading Tables 58
Comparing Percentages 60
Interpreting Relationships 66
Questions 68

7 **Measures of Association** 69
Logic of Measures of Association 69
Ordinal Measures 75
Nominal Measures 82
Interpreting Association 85
Questions 89

8 Statistical Controls 90

How to Identify Spurious Relationships 91

How to Interpret Controls 94

Other Ways to Look at Three-Variable Relationships 98

Use of Additional Controls 100

Questions 101

9 Changing Variables 103

Recoding 103

Indices and Scales 106

Summary 114

Questions 114

10 Statistical Inference 116

Probability Theory 116

The Central-Limit Theorem 123

Hypothesis Testing 126

Confidence Intervals 132

Directional Test 134

The t-Distribution 136

Summary 138

Questions 139

11 Interval Statistics 140

The Role of Interval Statistics 140

Correlation and Regression 142

Controls 154

Multiple Regression 157

Factor Analysis 163

Interval-Analysis Techniques 167

Questions 167

12 Relationships Between Nominal and Interval Variables 168

Comparing Two Means 168

Measuring Strength of Relationship 174

One-Way Analysis of Variance 179

Two-Way Analysis of Variance 185

Summary 194

Questions 195

13 Research Reports 196

Scope 196

Organization 198

Style 199

Summary 200

Question 200

Answers 201

Further Readings 206

Index to Notation and Statistics: Terms and Symbols 209

Index 211

Preface

Why be concerned with data analysis? Is it not a technical matter, one best left to mathematicians or statisticians? Perhaps that once was true, but such a view does not recognize the realities of modern society. Whether one be a doctor, lawyer, journalist, physicist, social scientist, or simply an informed citizen, one must be able to analyze and interpret data. For example, the legal profession is now using survey research and analysis to assist in making decisions on change of venue and jury selection. Data analysis is necessary at every level of the business world, from production and inventory decisions to market research and pricing. With the proliferation of small inexpensive microcomputers, large amounts of data are becoming available to even the smallest businesses, and with these data comes the need for careful analysis.

This book is intended to explain the general principles of data analysis so that the reader will be able to read reports based on the analysis of data and know how to analyze data. A full range of statistical techniques will be introduced, from the simple statistics for a single variable to more complex techniques for the analysis of many variables. The complex techniques include analysis of variance, causal analysis, and scaling.

To obtain experience with data analysis, the procedures described in this book can be used along with the SETUPS packages, which can be obtained from the American Political Science Association. Users of the SPSS statistical package would use the FREQUENCIES procedure for the statistics described in Chapter 5, the CROSSTABS procedure for Chapters 6–8, the RECODE procedure for Chapter 9, the ANOVA procedure for Chapter 10, PEARSON, CORR, REGRESSION, and FACTOR for Chapter 11, and BREAKDOWN for Chapter 12.

In using this book for a two-term course, Chapters 1–7 should be used for the first term and Chapters 8–13 for the second term.

We wish to thank Thad Brown and Richard Niemi for their helpful comments on the original manuscript. The Center for Political Studies data used in this book were provided by the Inter-university Consortium for Political and Social Research, which bears no responsibility for our interpretations.

January 1980

Bruce D. Bowen
Herbert F. Weisberg

Reference Tables

Table 2.1	Maximum Sampling Error for Samples of Various Sizes	16
Table 6.4	Maximal Sampling Errors for Differences in Proportions	63
Table 7.12	Chi-Square Values Required for Significance	87
Table 10.1	Areas Under the Normal Curve	122
Table 10.2	Error Conditions When Accepting or Rejecting a Null Hypothesis, Depending on Whether the Null Hypothesis Is True or False	129
Table 10.3	Critical Ratios for the *t*-Distribution	137
Table 11.3	Pearson *r* Values Required for Significance	152
Table 12.6	Distribution of *F*	182

To Bryan

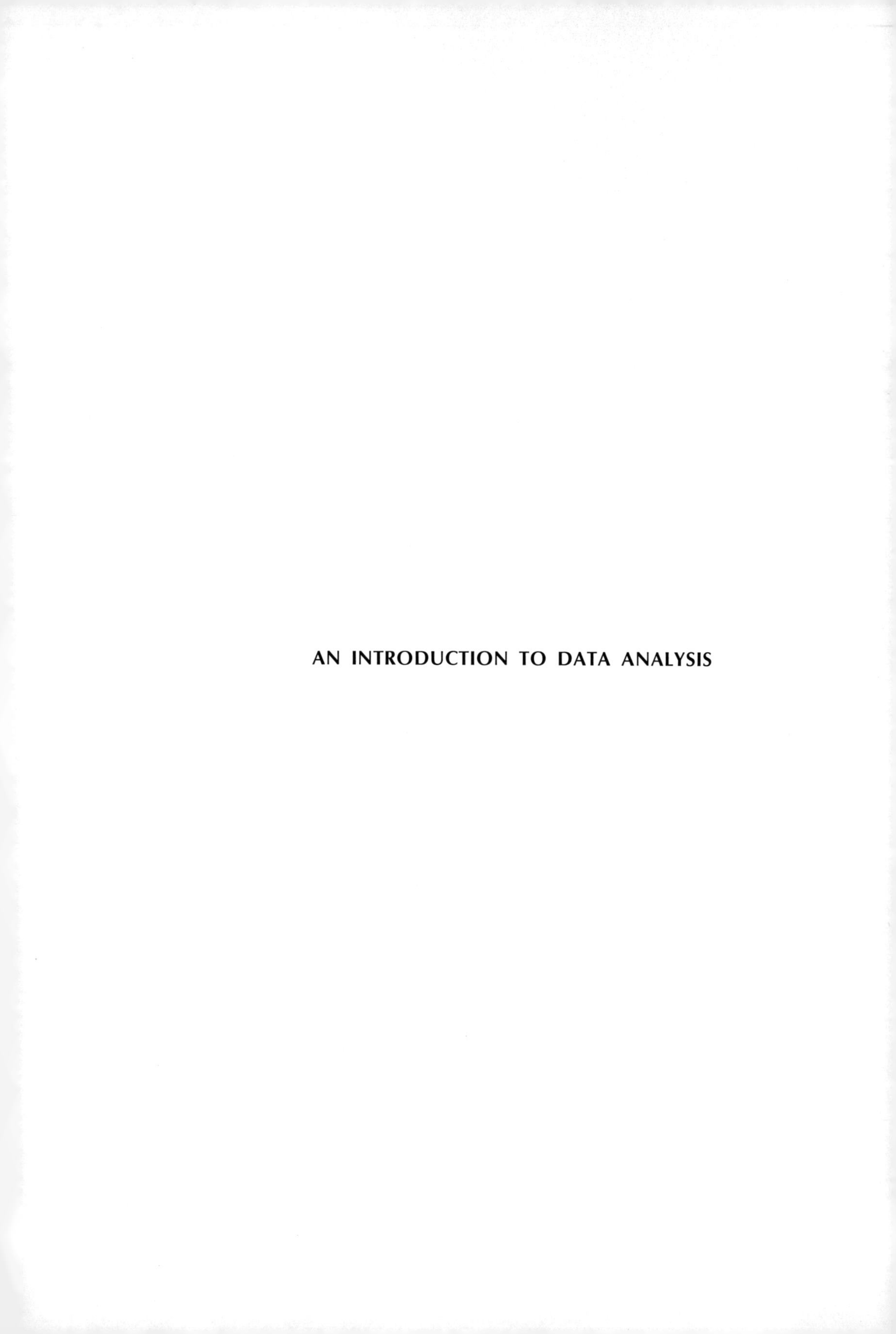

AN INTRODUCTION TO DATA ANALYSIS

1

The Role of Data Analysis

The definition of literacy has expanded in our modern world. The ability to read prose is no longer a sufficient skill for coping with our technological society. Instead we must now be able to read and make some sense of the mass of data that is collected and presented to us each year. Data are now everywhere, even in the social realm. Public-opinion polls gauge reactions to and evaluations of the president and alternative political candidates. Crime statistics indicate trends in the number of crimes reported and solved. Economic statistics detail every twist and turn in the nation's economic health. Government reports evaluate social programs, indicating statistically which have succeeded and which have failed. Data are too pervasive to leave their analysis to the technical mathematicians. Instead, social scientists, doctors, lawyers, business people, and citizens must be directly concerned with analyzing and understanding data.

In explaining data analysis in the following chapters, we present some necessary background information (such as explanations of the research process and computer utilization) and then progress from some of the simplest forms of data analysis to some of the most complex. In this first chapter, we begin with a general discussion of the role of data analysis.

THE IMPORTANCE OF DATA ANALYSIS

There are four reasons why understanding data analysis is important. An understanding of data analysis is required for (a) the intrepretation of research, (b) the evaluation of research, (c) clear thinking about empirical questions, and (d) planning research. We will now consider each of these in some detail.

Interpretation of Research

Suppliers of data generally provide their own interpretations. Pollsters headline who is ahead in the polls; the FBI explains crime trends; government reports indicate when the cost of a program outweighs its benefits. If we are willing to accept the interpretations of the groups producing the data, we need not come to grips with the analyses ourselves.

While the data themselves may be objective, the interpretation or explanation of those data is often not objective. In fact, data interpretation is often quite subjective. Two analysts can interpret the data in quite different ways. For example, two economists may use the same data to make two very different forecasts regarding the nation's economic future. It is far too simple to say that one of them is right and that the other is wrong, in an objective sense. They both may be right or wrong. They are probably looking at the data in quite different ways. As a result, it is important for citizens to be able to do some data analysis themselves so at least the conclusions of others can be evaluated.

Some people read only the words in reports about research and skip the tables and figures. Such an approach is too trusting, for it means that the reader will not notice if the data do not support the report's conclusions. Often the writer of the report is trying to justify his or her own personal views as to what should be done and therefore deliberately or accidentally interprets the data selectively so as to bolster his or her preferences. Citizens must be able to evaluate reports themselves if they are to determine what research actually has proved as opposed to what the author of the report wishes it proved.

Unfortunately, many of those who analyze data in the mass media (and elsewhere, for that matter) are not very careful about their analyses. They frequently draw conclusions that are not justified by the data. One common offender is the news commentator who offers some data (often a poll or survey) as evidence for a particular conclusion. All too frequently, the data do not support the commentator's conclusion. The commentator is sometimes spared considerable embarrassment in such situations because he or she is often correct for other reasons. However, as social scientists, we would like to rely more on careful data analysis than on luck (or even judgment) to keep us from being frequently embarrassed.

Evaluation of Research

Additionally, there may be problems with the research itself. For example, poll questions might be worded in such a way that they lead to an exaggeration of the phenomenon being measured. The wording of a question regarding support for the president might materially affect the measured level of support. Another example is crime statistics, which are notoriously weak because large number of crimes go unreported. As a consequence, it is important for citizens to understand the properties of high-quality research and data, so that they have standards with which to evaluate research.

When reading a newspaper headline on some research project, the natural tendency is to assume that the research was performed in a scientific manner. However, it is necessary to go beyond the headlines to determine how the research was conducted and whether it can be trusted. Some polling techniques give more accurate results than others, for example. Many polls conducted by newspapers are actually so informal that their results should not be believed at all. We would be more likely to believe a report evaluating a social program if it were based on an extensive study of the program rather than a quick look at its operation in a nonrepresentative community.

The implication of this point is that data analysis cannot be studied in isolation from the design of the research. To analyze data, it is vital to understand how the data were collected. Therefore, we review the basics of research design in the next chapter, as a preliminary to our presentation of data-analysis procedures.

Logic of Empirical Analysis

Rather than viewing data as imposing onerous demands on citizens, we feel that a familiarity with data analysis leads to important new perspectives on logical processes. Everybody talks about "causes" and "explanations," but most of us use these terms in quite informal ways. Confronting real data makes one think more systematically about causation and explanation. Instead of casually declaring something to be the cause of something else, one begins to think through what would be necessary to prove that causation is involved.

Perhaps the most important part of this process is the realization that empirical data may be required to test propositions about causation. We often talk as if causation were simple to discover, as if we could explain anything that happens. But a causal statement is actually a claim that one thing leads to another, and such a statement can be empirically false. Data analysis should make us sensitive to the importance of empirical information about claimed causal processes.

It also permits us to think about how to disprove something. Rather than accept a research report as proving a point beyond question, we can begin to think about how we could disprove that result in favor of some other result. Might some other research be able to test out an alternative explanation? If we distrust the conclusions in a research report, we can lay out logically how an entirely different process could have yielded the results in the first report. Thus data get us into new patterns of thinking about proof, causation, and empirical processes. With this in mind, in Chapter 3 we discuss the nature of causation and explanations in science.

Planning Research

Finally, the researcher setting out on a research project must understand in advance the nature of data analysis. One might imagine that data can be collected in any old way and then analyzed, but it is important to have the intended analysis in mind when collecting the data. Some types of analysis require certain types of data, so advanced planning is required to ensure that the data meet the requirements of the analysis techniques. The more thoroughly the researcher plans through the desired analysis before actually beginning the research, the more likely it is that the research will be designed to permit the desired analysis.

As an example, consider the decision as to the number of people to interview in a poll. In order to make that decision, it is necessary to understand how data analysis of polls is affected by the size of the sample. This is a technical question that requires an understanding of analysis procedures.

Additionally, it is necessary for the researcher to understand how computers are used in data analysis if the researcher is to set up the data properly for computer analysis. Computers are used extensively today in data analysis, so the person doing research should understand how computers would be used. Because of this consideration, we devote our final preliminary chapter, Chapter 4, to a description of the role of computers in data analysis.

WHAT IS DATA ANALYSIS?

We are concerned in this book with the procedures employed in the analysis of data. There are formal mathematical procedures for use in understanding data, procedures that are generally referred to as *statistics*. However, actual social-science data rarely satisfy the elegant assumptions of mathematical statistics. Therefore, we choose to call our concern in this book "data analysis" rather than "statistics." We may cover much of the same material that would be covered in a statistics course, but in a much less formal manner and with a greater sensitivity to the problems encountered with real data.

What do we include under the rubric of data analysis? The simplest form is the tabulation of results, such as counting the number of people giving each possible response to a survey question. This type of analysis is explained in Chapter 5. The next most complicated technique involves examining relationships, such as seeing whether children given some sort of "treatment" (such as preschool) perform better in school than those not given that "treatment." Chapters 6 and 7 focus on the analysis of such relationships, and we return to that topic for numerical data in Chapters 11 and 12. The most complicated forms of analysis involve looking at the relationships involving many topics at once, such as separating the effects of preschool from those of family background. Chapters 8 and 9 begin this topic, with further consideration in Chapters 11 and 12.

The final stage involved in data analysis is writing the results. This stage is critical, because the advancement of science requires the communication of the results of research. Therefore, our final chapter focuses on the writing of research reports.

2

A Review of Research Design

Data analysis depends very directly on many design considerations of the research. There are correct and incorrect ways to conduct a research project, and even high-quality data analysis cannot rectify fundamental errors in the research design. Therefore, when analyzing or evaluating the analysis of data, it is necessary to detect any research-design weaknesses that limit the data analysis. Consequently, we devote this chapter to a review of some key facets of research design.

THREE BASIC PROBLEMS OF SCIENTIFIC RESEARCH

Scientific research must confront three basic problems:[1] measurement, control, and representation. Each of these problems must be solved in a satisfactory manner in the design stage if the research is to lead to conclusions that can be generally accepted.

Measurement

The first requirement is high-quality measurement. In public-opinion polls, the measurement concern involves the wording of ques-

[1]Leslie Kish, "Some Statistical Problems in Research Design," *American Sociological Review, 24* (1959), pp. 328–338.

tions. Are the questions worded in a sloppy manner, or do they correctly assess the concepts that the researcher wants to measure? In a study of the effectiveness of an experimental teaching program, the measurement problem would be how effectiveness is measured. Are the tests used to measure success of the students really measuring that, or are they poor tests? We return shortly to a more detailed treatment of the components of good measurement.

Control

The second requirement of good research is effective control. If we are attempting to demonstrate that something has an effect, we are trying to show that a difference exists between when that thing is present and when it is absent. For example, showing that students do better on a test after exposure to an experimental teaching program proves nothing at all about the effectiveness of that teaching program, because they might have improved just as much (or even more) with a continuation of the existing teaching program. If an effect is to be demonstrated, the researcher must be able to control the situation sufficiently to give the treatment to one group but not to another. For example, if the students exposed to an experimental teaching program improved but an equivalent group of students not exposed to that program did not improve, then we would conclude the experimental program had an effect. When such "physical" control is not possible, "statistical" control may still be possible. For example, no researcher can control whether a person is male or female. But if it is thought that the different sexes would react differently to the experimental program, the researcher can compare men exposed to a program with men not exposed to it and similarly for women.

In some settings, the researcher can control the assignment of subjects to groups, setting up the design so that the groups given the experimental program and not given that program are originally equivalent. In other settings, the researcher cannot make the groups equivalent but at least can assign subjects to the groups. One effective procedure for doing this is *randomization.* If subjects are assigned to groups randomly, then those groups should differ in only random respects. This provides a means of controlling for a number of factors, and it is particularly useful when the researcher is unsure what factors should be controlled explicitly. Since unknown factors may always be present, randomization is generally preferable to non-random assignment of subjects to groups.

Representation

The control problem involved whether the finding for the people studied held true for those people. The representation question asks whether we can generalize from the people studied to some larger population that interests us more. We may interview 1,500 people in a poll, but we are interested in generalizing about the entire population (of the country, state, or city we are studying) and not just those 1,500 people. We may study the effectiveness of a teaching program on classes in one school, but we want to make generalizations about the effectiveness of the teaching program which are not limited to that one school.

The representation question depends on how the people studied are selected. Generalization is possible if the people are representative of a larger group, and that holds if they are selected in such a way as to ensure that they are representative. If the people are selected in a nonrepresentative manner, generalization will not be possible. Thus, the representation problem leads us into the question of how the people studied were chosen. Later in this chapter we devote several pages to a detailed consideration of sampling issues related to representation.

THREE TYPES OF STUDIES

There are three basic types of studies:[2] the experiment, the sample, and the use of unobtrusive data. These types of studies differ in how well they handle the research problems just explained.

Experiments

Experiments are a useful way of studying change. For example, if the effects of communications on attitudes were to be studied, an

---

[2]Leslie Kish, "Some Statistical Problems in Research Design," *American Sociological Review, 24* (1959).

experiment could perform the following steps. First the attitudes of all of the subjects would be measured. Then each subject would be randomly assigned to one of the two groups—one group to receive the experimental communications ("experimental group") and one not to receive them ("control group"). Next, the experimental communications would be transmitted to the first group. Finally, the attitudes of all the subjects would be remeasured, and the attitude changes for the two groups would be compared.

Experiments excel in control because, as in the example above, one can be fairly certain that no extraneous factors caused the observed changes. In fact, many experiments are conducted in laboratories where the subjects are isolated physically from extraneous factors that might affect the results.

The difficulty with experiments is that they are weak on representation. For one thing, the laboratory conditions are highly artificial. Communications may have effects in the experimental laboratory that are different from effects they would have in natural settings. The setting is not representative of natural situations, so it is unclear whether one can generalize the results to real situations. Also, the people studied in experiments are often not representative of the population of true interest. Most attitude-change experiments are conducted on college students. However, the researcher usually wishes to make generalizations about how people generally are affected by communications, not just about how college students are affected. To make this argument, the researcher would have to claim that age and amount of education did not affect susceptibility to attitude change and that the college students were behaving naturally in the experiment rather than trying to please (or screw up) the researcher. Because experiments are rarely performed on the population of true interest, generalization from experiments can be hazardous.

Samples

A second basic type of study is the sample. This type includes sample surveys, where a sample of people are asked a set of questions, but it is not limited to surveys. For example, we would include here studies of newspapers for a sample of days.

Samples excel in their representation. They are generally conducted in the natural setting, as opposed to the experimenter's artificial laboratory. And the sampling procedure can ensure that the sample is scientifically representative of the larger population of interest, as by randomly choosing people to interview.

However, samples are generally weak on control. The researcher cannot control whether or not a treatment is given to a person being interviewed. At best, the researcher can examine whether people in one category differ from people in another category, but that can lead to inference problems. For example, say that a researcher interested in the effects of television news finds that those who watch television news are more informed than those who do not watch it. If an experimenter had controlled who had to watch television news and who was not permitted to watch it, the researcher could conclude that differences in information were due to watching the television news. Unfortunately, such control is not possible in surveys. The survey researcher simply knows that the people who watched television news are more informed than those who do not watch it. The reason might be that people who are more informed are more likely to choose to watch television news than are those who are uninformed. The survey researcher lacks control over the assignment of people to groups, so determination of which is the cause and which is the effect is impossible in this example.

Unobtrusive Data

There are many other methods of investigation, such as using available official data. The analysis of legislative roll call votes falls in this category, along with the analyses of census data, historical election returns, statistics relating to the severity of wars, and so on.

The advantage of these types of data is that they are collected for reasons having nothing to do with the researcher, so one need not worry about experimental subjects or survey respondents modifying their behavior or attitudes to try to please the researchers.

The prime difficulty with such data involves measurement. The researcher did not control how the data were collected, so the available data may not correspond precisely to what is needed to test the researcher's theories. The researcher will have to make do with what is available rather than tailoring the data collection to suit the research questions.

Additionally, there can be problems with both control and representation. The researcher is at the mercy of whatever process generated the data, without being able to assign cases to groups randomly or being able to ensure that data are available for a representative group of cases. In some research settings, this may not be a problem, but other research on available data suffers considerably from the researcher's lack of ability to manipulate the data-collection process.

THE QUALITY OF THE MEASUREMENTS

In any study, the quality of the measurements must be considered. We review in this section a series of aspects of good measurement.

Reliability

First, good measurements must be reliable. Imagine measuring the length of a pencil with an elastic ruler. One person may measure its length to be 3 inches, while another person may find its length to be 4 inches because she stretched the ruler less. This is an example of nonreliable measurement. In a reliable measurement, repeated measurement of the same thing would give very similar results, as is the case when measuring the length of a pencil using a wooden ruler.

An example of a nonreliable question in a survey is the question, "Do you favor reducing American use of gasoline by increasing our taxes on foreign oil?" A person who wants to increase taxes on foreign oil but does not want to reduce our consumption of gasoline would have difficulty responding. He might answer yes one time and no the next, just because he is focusing on different parts of the question each time.

Reliability problems also arise when information is coded from verbal material. Say that the researcher has hired people to read political speeches and code them as to whether or not they make ideological appeals. Different coders might interpret the term "ideological appeals" differently, so that the coders' opinions would affect the research results as much as the actual content of the speeches would. In such studies, the researchers must ensure that the interpretations of the coders are uniform.

Validity

A second measurement problem is validity, whether the measure actually measures the concept of interest. Sometimes the measure may actually be measuring a related concept rather than the one the researcher wanted to measure. Say a researcher was interested in the effects of having children in school on a person's support for increased school taxes. The survey question "How many children do you have?" may be reliable, but it is not a valid measure of how many children they have in school. Given their objectives, the researchers would want to ask, "Do you have any children in public school?"

The validity problem is especially severe for the researcher using available data. The available measures may not correspond exactly to what the researcher wants to measure, but there may be no choice but to compromise. After all, the researcher does not control what data the government collects (or especially what data the government collected a century ago). When this happens, the researcher should candidly admit that data-availability problems led to modifications in what was studied.

Unbiasedness

Additionally, good measurements are not biased. In a biased measurement, the researcher makes certain results more likely. An example of a biased survey question is "Should we continue disarmament so as to eventually turn over this country to the Communists?" Such a question would find less support for disarmament than would a neutrally worded question. In our previous example of coding speeches as to whether or not they make ideological appeals, one could imagine an ideological researcher giving instructions to coders that bias the results, as if a conservative researcher considered liberal arguments to be ideological but did not consider conservative arguments as ideological.

SAMPLING PROCEDURES

Our final topic in this chapter is sampling procedures. Sometimes it is possible to study every case of interest, but often the cost would be prohibitive, so sampling is substituted. Instead of interviewing 80 million voters in the United States, pollsters talk to a few thousand scientifically selected respondents. Rather than analyze every campaign speech given in an election year, the researcher may sample a manageable number of the speeches.

Sampling is widely used in the sciences and in technology, and a relatively complete body of mathematical and statistical theory has been developed to guide its application. Correct sampling procedures require very strict adherence to certain logical principles, but they are not intrinsically difficult to understand. In this section, we shall describe some of the sampling procedures that pertain to the social sciences.

Types of Samples

The most important distinction is between nonscientific and scientific sampling. In nonscientific sampling (nonprobability sampling), cases are studied because they are considered "typical," because they are readily available, or because the researcher wants to study them. When a professor conducts research on her own students, the sampling is nonscientific. Because of that, it is not clear how the results would generalize to the total population of students.

The contrast is with scientific sampling (probability sampling), where each person in the population has a known (and generally equal) probability of being included in the sample. A random sample of voters drawn from the voter-registration lists is an example of a scientific sample.

There are actually several techniques for scientific sampling. The purest form is known as the "simple random sample." The researcher would use a list of the population and would give every combination of people an equal chance of being selected. In principle, this could be done by writing each name on a piece of paper, putting all the names in a hat, mixing up the names, and drawing the sample.[3] For a large population, the procedure would no doubt be automated by using a computer to select the names from the list. Simple random samples are representative of the population from which they are drawn, but they are expensive, because they require a listing of the entire population and because the separate cases included in the sample may be very physically isolated. Simple random samples of the U.S. population are impossible, because there is no accurate listing of every person and because such a listing would be prohibitively expensive to produce. Also, interview studies could not afford to fly an interviewer into

[3]Note that the term *random* does not mean "haphazard selection," as it sometimes does in everyday usage. It has a very precise scientific meaning here, based on the notion of each possible sample having a known probability of being selected.

Snowflake, Arizona, for the one interview to be taken there and then on to Casper, Wyoming, for the next interview. The transportation costs for interviews require that several interviews be physically clustered near one another.

As a result of these cost considerations, scientific samples often modify the simple random sample. They might, for example, cluster sample—conduct several interviews on the same block so as to minimize transportation costs. Such sampling remains scientific so long as the people interviewed are chosen in a nonsystematic manner (not letting the interviewer choose whom she would feel most comfortable interviewing). And such sampling can be more accurate than simple random sampling, because the reduced cost of clustering might permit so many more interviews to be taken that the cluster sample is more accurate at a fixed cost than would be the smaller simple random sample at that cost. Most national samples used in public-opinion surveys are scientific samples, but they involve complicated variations on the simple random sample.

Sampling Error

There are inevitable difficulties associated with sampling, the most important of which is termed "sampling error." Sampling error is the error that arises from trying to represent a population with a sample. Almost inevitably, the sample differs in some way from the population—at least, one has to take into account the possibility of such a difference. The consequence is that we should not take sample results as absolutes but as approximations.

The sampling error is based on a number of considerations, including how the sample was selected. It is impossible to calculate the sampling error in nonscientific sampling; the advantage of scientific sampling is that the sampling error can be computed. For example, in a national survey of about 1,500 people using common scientific sampling procedures, the sampling error is about 3 percent. If the survey finds that 43 percent of the sample favor legalizing marijuana, what that really would mean is that the odds are very high that the proportion of the population favoring legalizing marijuana is between 40 percent (43 − 3) and 46 percent (43 + 3). This is actually a probability statement. More precisely, it means that in 95 out of 100 samples (which is considered a high and safe probability) the population proportion would be within 3 percent of the sample proportion. These topics are treated in considerably more detail in Chapter 10.

Obviously, the smaller the sampling error, the more precise we can make statements. We would prefer 3 percent sampling error to 20 percent, because to say that support for legalization of marijuana is between 23 percent and 63 percent would be so broad as to be meaningless. So how is the sampling error reduced?

Sample size. The key ingredient in determining the sampling error for scientific samples is the absolute sample size. The more cases in the sample, the smaller the error. Actually, quadrupling the sample (multiplying the sample size by 4) cuts the error rate in half. If the sampling error with 1,500 interviews were 3 percent, that sampling error could be reduced to 1.5 percent by instead using 6,000 interviews. Most surveys stay with 3 percent sampling error, because the extra expense of 4,500 more interviews is not worthwhile if sampling error is reduced *only by 1.5 points.* Rarely do we have to be within 1 or 2 percent of the true value. Also, there are other errors in surveys—such as errors in reading the questions and errors in recording the answers—that cannot be eliminated entirely, so an extra 1.5 percent sampling precision would not be worth many thousands of dollars given the other unknown errors.

The sampling fraction. The sampling error also is affected by the sampling fraction—the percentage of the population that is being studied. This is important when the sampling fraction is so large that the sample must be fairly similar to the population. Consequently, when the sampling fraction is large (say, 30 percent or larger), the sampling error is less than it would have been for the sample of the same size from a larger population. However, sampling fraction is too small to matter when "only" a few hundred or thousand cases are studied from a population of several million. Thus, the sample size always matters, but often the sampling fraction does not. For example, for the same accuracy we would need the same size sample to represent voters in Peoria, Illinois, as to represent voters in the entire United States.

Assumptions

Conventional sampling-error calculations (and most forms of statistical inference) are usually based on two assumptions: (a) that sampling has been employed rather than studying the entire population and (b) that the sample is a simple random sample. Often those assumptions are not met. If a researcher is studying an entire population (such as all members of an eleven-person city council), then by definition there is no sampling error. While the issue is complex, studies that quote sampling errors when studying an entire population are overstating their claim to scientific status. If scientific sampling is employed, but the sample is not a simple random sample, then more appropriate sampling-error formulas can be obtained with the aid of a statistician. As a rough guide, the sampling error for a typical national sample might be one-third higher than the simple random sample formulas would indicate. Table 2.1 shows the sampling errors for various-sized samples under simple random sampling and a typical complex national survey taken by the University of Michigan Survey Research Center.

Finally, sampling errors and statistical inferences focus attention on whether a finding is likely to hold in the underlying population rather than on whether a finding has any real substantive importance. With a large-enough sample, any finding would be statistically significant, even if it has no substantive importance. For example, a preelection

Table 2.1 Maximum Sampling Error for Samples of Various Sizes

Sample Size	Simple Random Sample (%)	Survey Research Center Survey (%)
2,000	2.2	3
1,500	2.6	
1,000	3.2	4
700	3.8	5
600	4.1	
500	4.5	6
400	5.0	
300	5.8	8
200	7.2	
100	10.3	14
90	10.3	
80	11.0	
70	11.7	
60	12.7	
50	13.9	

NOTE: These are maximum sampling errors, because sampling errors depend on the proportion being estimated. Sampling errors are maximal in estimating proportions around 50 percent. There is less error in estimating proportions less than 30 percent or above 70 percent particularly in estimating proportions less than 10 percent or above 90 percent. Yet, in any event, the sampling errors are not greater than those shown in the table.

SOURCE: The Survey Research Center values are taken from Leslie Kish, *Survey Sampling* (New York: Wiley, 1965), Table 14.1.I, p. 576. They are based on the 1963 Survey of Consumer Finances and may differ from survey to survey. Reprinted by permission of the publisher.

survey of enough voters might be able to proclaim that the Democratic candidate for president is one-tenth of a percentage point ahead of the Republican candidate. But even if the sampling error is so small to permit such an assertion, the Democratic lead is too shaky to be taken seriously. What is important in this instance is that the election is close, not that the Democrat is ahead.

SUMMARY

Any serious research must deal with the problems of measurement, control, and representation. Quality measurement requires being able to obtain very similar results on repeated measurement (reliability), measuring the concept of interest rather than a related concept (validity), and not doing the measurement in such a way as to make one outcome of the research particularly likely (unbiasedness). Control is achieved in measuring the effect of some treatment by exposing one group to that treatment while not exposing another equivalent (control) group. Representation deals with whether the people studied are representative of the larger group of interest, so that the results of the study can be generalized beyond the people studied. Scientific sampling procedures best guarantee that the people studied are representative of the population of interest.

Questions

Critique the following newspaper reports on the results of research studies.

1. "A study has found that 2 percent of disco dancers get cancer. Therefore, disco dancing is dangerous to one's health." What is the control problem in this study?
2. "A study of undergraduate students at Yale University shows that nearly all have traveled abroad. This shows that the younger generation is traveling abroad in high proportions nowadays." What is the representation problem in this study?
3. "A study has found that northern states have higher tax rates than southern states. This shows that northern states are spending more money on social services than are southern states." What is the measurement problem in this study?
4. "A survey has found that 52 percent of the public back the incumbent governor's performance in office. Therefore, he still has majority support in this state." What is the sampling problem in this study?

3

The Analysis Process

The analysis of the data produced by a large research project can take years. It is a phase of research that requires careful planning and cannot be dismissed as a simple process. This chapter will introduce the general approach to data analysis taken in the social sciences and will discuss the nature of explanation and the design of an analysis effort. Later chapters will focus on the various procedures that are available for statistical analysis of data.

EXPLANATION

In social-science research, we are fundamentally concerned with explanation. We seek explanations for people's behavior and their attitudes. This does not mean that we are interested in explanations of an idiosyncratic action. Instead, we seek to test generalizations that account for the attitudes and behaviors of aggregates of individuals—generalizations derived from general theories of human behavior. Thus, we would not be concerned with why one particular person voted Democratic in 1980, but we might examine the generalization that unemployed people tend to vote against the incumbent administration.

Note also that real progress is made by a process of elimination—by disproving alternative hypotheses rather than by simply gathering evidence for one hypothesis. A set of data cannot *prove* an hypothesis to be true, since some later data may contradict that hypothesis. However, data can be used to disprove an hypothesis. If we try to think of all the plausible ways by which some result could have been produced and if we use the data to disprove all but one of them, that constitutes support for the remaining hypothesis. In line with this logic, we may speak of "disproving" an hypothesis, but we will not speak of "proving" an hypothesis to be true.

Terminology

Cause. In seeking explanations, social science is attempting to find the fundamental causes of social phenomena. But how are we using the term "cause"? The term has a number of meanings, but in scientific usage, its meaning is similar to that of "producing."[1] Something is a cause of another thing if a change in the first *produces* a change in the second. By contrast, we shall speak of two things as being "related," "correlated," or "associated" if changes in the two tend to accompany one another, and we shall sometimes speak of one thing "affecting" another if we do not wish to claim that the relationship between them is causal. Correlation alone does not prove causation. Indeed, causation is impossible to verify but easy to refute. Therefore, social scientists must be content with determining that two things are related to one another. For example, if we cannot prove that becoming unemployed causes a person to vote against the incumbent political party, we can at least test whether there is a tendency for these events to be associated with one another.

Variables. At this point it is useful to introduce some more terminology that we shall use in later chapters. A *variable* is, according to the standard dictionary definition, something that varies. Variables in social science include behaviors, attitudes, background characteristics, and so on. The variable that we wish to explain in a given study is called the *dependent variable.* For example, many studies of voting behavior seek to explain why some people vote Republican and others Democratic. In those studies, voting is the dependent variable. The other variables used in the explanation are called the *independent*

---

[1]Mario Bunge, *Causality* (Cambridge: Harvard University Press, 1959), pp. 46–48; Hubert M. Blalock, Jr., *Casual Inferences in Nonexperimental Research* (Chapel Hill: University of North Carolina Press, 1964), p. 9.

variables. The latter are also called the *predictors* of the dependent variable and are said to predict or explain the dependent variable. In our example, such independent variables as religion and race might be used to help explain a person's vote. The distinction to keep in mind is that a dependent variable is an effect, and an independent variable is a cause—or at least a suspected cause.

There are a few variables that social scientists frequently study that are almost always independent. These are demographic variables, such as race, region, sex, and age. Social scientists are seldom interested in the causes of such demographic variables and thus would have little reason to use them as dependent variables. There are also some variables that are nearly always dependent. These are usually behavioral variables, such as voter turnout, vote choice, and political participation. There are many other variables (especially attitudinal variables) that may be independent or dependent, depending on the context in which they are studied. For example, party identification (that is, whether the person thinks of himself as a Republican or an Independent or a Democrat) would be a dependent variable if we were studying the effect of income on party identification, but it would be an independent variable if we were analyzing its effect on a person's vote.

Aspects of Causation

As we have said, it is impossible to verify that an independent variable causes a dependent variable; however, there are three important tests of a relationship that are necessary parts of causation. If any of the three is not satisfied, then we do not have a causal relationship.

Association. First, causation can be said to exist only if there is some tendency for a change in one variable to result in a change in the other. Statistics can be useful in testing whether variables are correlated. However, we must recognize the possibility of complicated forms of association. In the simplest form of association, a unit change in one variable would always result in a uniform change in the other variable; for example, such an association would exist if each additional year of education always resulted in an average increase of $800 in a person's annual income. Of course, such simplicity is hard to find in human society. A more complex possibility is that an increase from one year to two years of education may produce an average increase of only $200 in annual income, and an increase from twelve to thirteen years may produce an average increase of $1,400 in annual income. Because the associations between such variables may be quite complex, we require a variety of statistical analysis procedures to permit us to test complex models of social behavior.

Temporal Order. Second, we can speak of causation only when the cause precedes the effect. Temporal succession of cause and effect is implicit in the notion of causation. When we use surveys, our measurements of both variables are usually simultaneous; we often examine the relationship of people's responses to two questions in the same survey. Even so, it is often clear that one thing has preceded the other. If we ask people how many years they went to school and their annual income, for example, we can safely assume, for most people, that their formal education preceded the attainment of their current income level.

Occasionally, it may seem that each of two variables affects the other. Usually, in such cases of "reciprocal causation," there is a clear back-and-forth interplay: a change in the first variable produces a change in the second, which, in turn, produces a change in the first at a still later time, and so on. Unfortunately, the available data may not permit disentanglement of this causal process. For example, there is a sense in which a vote may not only be affected by party identification but may itself affect party identification, as when someone who has always considered herself a Democrat finds herself voting Republican in a series of elections and then begins thinking of herself as a Republican. Without repeated surveys, it might be difficult to assess the importance of both causal directions.

Alternative Causes. Third, before we can attribute an effect to a particular cause, we must eliminate alternative causal explanations. As researchers, we must posit reasonable alternative hypotheses and disprove them in order to claim that a cause has been found. Although we may strongly suspect that one variable is affecting another, both may in fact be determined by a third variable. Does a person's education cause that person's income, or does his or her family's social class cause both without education having an independent effect on income? Statistical procedures can help analyze this problem, but the difficulty is discovering which possible third variable might be affecting the two-variable relationship. An important role of theory is to help us decide which third variables to test as alternative causes. However, because we cannot investigate the enormous number of possible third variables, we can never prove—in an absolute sense—that a relationship is causal.

To summarize, testing of a causal explanation requires (1) checking to see whether the variables are correlated, (2) verifying that the presumed cause precedes the presumed effect, and (3) eliminating alternative explanations. We may not be able to prove causation in the social sciences, but we can test causal explanations and thereby develop a better understanding of social phenomena.

CAUSAL PROCESSES

Beyond appreciating the logical difficulty of proving causation, it is important to understand the possible complexity of causation. It is unlikely that a single independent variable will account fully for our dependent variable. Human attitudes and behavior are too complex to employ such a simplistic model. Usually we should be looking for a causal *process* rather than a simple cause.

Multiple and Indirect Causes

Most dependent variables in the social sciences are complex enough to involve multiple causes, each of which has some effect on the variable. There are statistical procedures that can be used to test the relative importance of independent variables, and different procedures can be used to check different assumptions as to how several independent variables interact. Even when we take into account the possibility of multiple causation, we should not expect a complete explanation of social-science variables. We will still be limited to making statistical generalizations that are better suited to predicting aggregate behavior rather than the behavior of specific individuals. Additionally, the inevitable errors in measurement of the variables (whether in surveys or other data-collection procedures) reduce our ability to make explanations. There is a random component to social-science variables, and we must be satisfied if we can explain up to the limit set by that random component.

Another problem for social scientists is that the very same phenomena can be explained in different ways, through different theoretical perspectives. A particular behavior that a political scientist would understand in one way would be understood by a psychologist in another way and by a social biologist in yet another way. Each explanation can be correct within each discipline's own framework.

Furthermore, causation can be indirect. An independent variable may affect the dependent variable through another variable. In studying causation, we might wish to make a distinction between (a) which variable has a greater *direct* effect on the dependent variable and (b) which variable has a greater *total* effect, including its indirect effect.

Arrow Diagrams

Causal processes are often represented by arrow diagrams such as Figure 3.1. An arrow means that one variable is thought to affect the other; the direction of the arrow shows the presumed direction of influence. Such arrow diagrams are generally read from left to right and from top to bottom.

Figure 3.1 shows an example of a causal process by which some theories explain voting behavior. Setting aside the substantive features of these theories, which are not our present concern, let us examine the usefulness of such arrow diagrams in depicting a theoretical model. In concise graphic form, the diagram depicts a complex process. Evaluation of the candidates is portrayed as a cause of voting choice, but

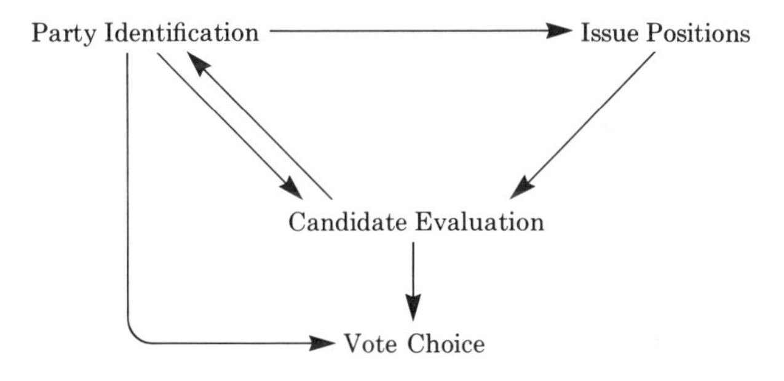

Figure 3.1
A Causal Model of Voting
(Adapted from John E. Jackson, "Issues, Party Choices, and Presidential Votes," *American Journal of Political Science, 19,* 1975, Fig. 1, p. 163.)

candidate evaluation is also shown to have its own causes. Some causes are shown to be indirect, such as issues affecting the vote through their effect on the evaluation of candidates. Party identification is shown to affect the vote directly and also indirectly, through evaluations of the candidates. There is an allowance for reciprocal causation with evaluation of the candidates shown to have an effect on party identification as well as vice versa. The statistical procedures used to test this model are quite sophisticated and beyond the compass of this discussion, but this example illustrates the usefulness of arrow diagrams in depicting the complexity of causation.

When planning an analysis, it is often useful to draw such arrow diagrams. This exercise forces us to think directly in causal terms, to consider the full range of variables that should be included in an explanation, and to map out the presumed causal process.

PLANNING AN ANALYSIS

The analysis of a set of data is never totally systematic. As researchers obtain their results, they get new ideas for further analysis. However, it helps to plan the analysis in advance as much as possible. Doing so may alert the researcher to potential analysis problems before a large amount of meaningless work is done.

Secondary Analysis

As we turn to the design of a data-analysis project, we must introduce a distinction between analysis by the original investigators and *secondary analysis* by others. It is becoming common for those who design the research project (the original investigators) to make their data available for secondary analysis by other researchers after they have analyzed it. The reasons for this development are largely financial. Large-scale significant research projects are quite expensive. Whether the project requires coding a large number of event interactions between nations, collecting data on the sentencing behavior of judges, or conducting a national survey, it makes good economic sense to share the data that is collected so that others need not duplicate the effort. It would be silly, for example, for there to be several academic surveys after each election, all seeking essentially the same data. Consequently, most major research projects release their data to the rest of the academic community after they have completed their reports.

There are now several major archives that store data sets released by investigators. For example, the Inter-University Consortium for Political and Social Research, based at the University of Michigan, has an extensive archive that contains major national-election surveys from the United States and many other countries plus a large number of surveys on other topics. In addition to survey data, the consortium also stores data on national attributes, U.S. Census data, data on world-event interactions, and data on many other topics. Often within a year of a survey, universities that belong to the consortium can obtain the survey data and non-member universities can purchase individual sets of data for a fee. The Gallup, Harris, and Roper polls have similar services that permit their surveys to be acquired by interested researchers.[2] Since these archives contain data from surveys done as long as thirty years ago, they permit researchers to evaluate gross attitude and demographic changes over time.

As a result of the development of such archives, secondary analysis is now very common—probably even more common than primary analysis. Most large universities have a large collection of survey and other data, with associated codebooks, available for secondary analysis by faculty and students. Unfortunately, the secondary analyst faces some special problems in the design of the data analysis, as we point out in the remainder of this chapter.

Analysis Design

It is important to design the data analysis with care rather than conduct an unsystematic analysis. "Analysis design" consists of three steps: the specification of the hypotheses to be tested, the choice of the specific variables to use in those tests, and the selection of appropriate statistical procedures to analyze the data. These steps must be carefully meshed. Variables must be found that properly operationalize

[2] A valuable collection of Gallup Poll results over the years is available in George Gallup (Ed.), *The Gallup Poll: Public Opinion, 1935–1971* (New York: Random House, 1972).

the concepts in the hypotheses, and statistical techniques must be selected that are appropriate for the data.

Hypotheses. The statement of hypotheses by primary investigators begins in the research-design phase with a statement of the study objectives. Since careful testing of hypotheses requires disproof of alternative explanations, the analysis phase should begin with a statement of specific hypotheses and alternative hypotheses. The task for the secondary analyst is very similar, translating his or her reasons for studying the problem into specific hypotheses for testing along with alternative hypotheses that test competing explanations.

Variables. Next is the choice of variables to be used in the tests. When analyzing surveys, questions must be selected that correspond as closely as possible to the theoretical concepts in the hypotheses. The original investigator takes care of most of this problem in the writing of the questions, though he or she may still have to combine some questions. Say, for example, that the hypothesis suggests that the amount of information a person has about a political campaign affects the person's vote. The original investigator might combine separate questions about how much the person has followed the campaign by television, radio, newspapers, and magazines into a general measure of campaign information. Procedures for changing measures and building new measures will be detailed in Chapter 9.

The choice of questions is much more difficult for the secondary analyst. Researchers approach problems from different perspectives and with different philosophies and ideologies, so the secondary analyst is not likely to be satisfied completely with the original investigator's questions. The secondary analyst is inevitably interested in hypotheses different from those of the original investigators, and so finds that some needed questions are either missing or not asked in the most desirable way. Therefore, secondary analysis often requires compromises, using the available questions that come as close as possible to those desired. In addition, the secondary analyst may have to use several surveys together, testing with one study the hypotheses that cannot be adequately tested with another study.

Statistical Procedures. Once the variables are selected, the analyst (primary or secondary) must decide which statistical operations are required to test the hypotheses. This usually begins with counting and "percentagizing" the responses to single questions (see

Chapter 5), but the main work is examining the relationships between pairs of variables (Chapters 6, 7, and 11), often with tests to see whether the two-variable relationships are different for people who differ on other variables (Chapters 8 and 11). This latter procedure is the basic approach that is used to test alternative hypotheses, so it is one of the most important parts of an analysis. Later chapters will describe these procedures in detail, but it should be realized that the steps described in Chapters 5–7 are largely preparatory for the *three-variable analysis* of Chapter 8. The selection of the statistical technique also depends on how the variables were measured, a consideration which will be explained in Chapter 5.

SUMMARY

This chapter has covered a variety of material concerning the nature of social-science data analysis. The analysis is intended to derive explanations for attitudes and behavior from raw data. Such explanations embody complex views of causal processes. An effective analysis requires careful preliminary planning.

Questions

1. Construct a model of a causal process underlying a person's income level. Use age, education, and parent's social class as predictors. What other explanatory variables might be useful? (To avoid some definitional problems, assume a restriction to people over 25 years of age.)
2. Construct a model of the causal process underlying a person's party preference. As predictors, use parent's social class, parent's party preference, and the person's social class. What other predictors might be included in the model?
3. Construct a model of the causal process underlying a person's attitude on abortions. Include in the explanatory variables the person's religion, the person's age, and the extent to which the person has a traditional value system. Consider what additional predictors might be included.

4

Computer Utilization

Social-science data analysis is generally performed with computers. Although this is not the place to attempt a thorough description of computers or even to survey the use of computers in the social sciences, we shall at least try to remove some of the mystery that still surrounds computers by describing the way they are used in the data-analysis process. Really, computers cannot do anything that could not, theoretically, be done with pencil and paper. They are used mainly because they save time—they can process huge amounts of data quickly.

DATA STORAGE

The typical research project produces a huge amount of data. As an example, the 1972 American election survey of the Center for Political Studies measured about 1,100 variables for 2,700 respondents and generated a total of nearly 3 million units of information. The sheer volume of this information would be impossible to process without a computer, at least within reasonable time and cost limits.

Imagine analysis for this survey without a computer. To find out how many people gave each response to a question, several clerks could sort the questionnaires into piles. The clerks would look through each questionnaire to find the question and response and then decide which pile it goes in. This might take a few hours; it would take just seconds on a computer. A two-variable relationship would be even more difficult to analyze without a computer. Clerks could divide the 2,700 questionnaires into piles depending upon the party identification and vote of each respondent. Several clerks would have to work for days to determine one such relationship; a computer could do so in seconds. We should bear in mind that the computer does not do anything that humans cannot do but that the computer can handle huge amounts of data much faster than humans can.

The first requirement for computer analysis is that the data must be entered into the computer. In the coding process, the coder assigns a numeric code to each response on each question. For example, the survey question regarding presidential vote might be coded by giving Republican voters the code number 1; Democratic voters, 2; other voters, 3; "don't know," 8; not ascertained, 9. Presidential vote would be just one of many coded variables, each with its own numeric code. Once the responses are coded into numbers in this manner, the problem of getting the numbers into the computer remains. To do this, the coded data are punched into computer cards that the computer can read.

Computer Cards

There are eighty columns on a computer card (Figure 4.1), which means that eighty one-digit numbers can be punched into each card. For example, presidential vote, which can be expressed by one digit, might be put in column 70; years of education, which requires two digits, might be put in columns 71 and 72. More than one card would be needed for each respondent if more than eighty columns were needed per respondent.

There are ten numeric rows (0 through 9) intersecting each column. A person who voted Democratic would have a hole punched in row 2 of column 70 of the card. If there are more than ten categories for a variable—such as the number of years of a person's education—two columns would be used together. If a person had fifteen years of education, then row 1 of column 71 would be punched along with the row 5 of column 72.

Figure 4.1 shows an actual computer card with a "2" punched in column 70 and a "15" punched in columns 71–72. Note that there are a total of twelve rows in each column; the top two rows are non-numeric codes which are usually not used in punching data (except for using one of them for the minus sign).[1]

Keypunching

The process of punching the cards is called *keypunching,* and the machine which does this is called a "keypunch." A keypunch is slightly larger than a typewriter with a typewriter stand. It has a keyboard similar to a typewriter's, but instead of printing on a piece of paper it simultaneously punches holes in the card and prints the number at the top of the column of the card.[2]

Keypunching is a point at which errors can creep into research. Since keypunching thousands of numbers per hour can be very boring work, the operator will make errors from time to time. Thus it is essential to check the keypunching for errors; this is done in a special process known as *verifying.* The punched cards are fed through a machine that looks almost exactly like a keypunch (but called a "verifier"). As each card comes into position, the operator of the verifier retypes the data for that card on the verifier's keyboard. Instead of punching new holes in the card, the machine verifies that there are holes where the operator indicates there should be holes. If there is a hole in the right place, everything proceeds. However, if the operator indicates that there should be a "7" in column 66 where there is now a "5," the verifier signals the error by stopping. Thus, the verifying process provides a good chance of catching keypunching errors. But note that every card is in effect being punched twice, which doubles the cost of the keypunching process.

Alternative Storage Modes

Putting the data onto cards is not enough. Consider, for example, the 1972 election study that used twenty-five cards for each of about 2,700 respondents—or a total of about 67,500 cards. So many cards are bulky to store and heavy to carry; it is easy to lose a few cards or

[1]Those top rows are used when punching letters and special symbols onto the card. Two or three punches in a single column are used in combination to denote a letter or symbol.

[2]The computer reads only the punches in the card—the printing is to allow humans to read at a glance what is punched in the card.

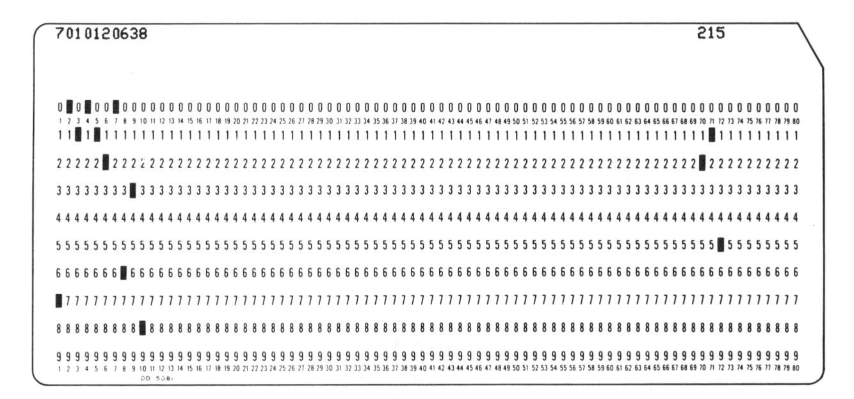

Figure 4.1
A Computer Card

get a few out of order. Worst of all, the cards are likely to start to warp after a few months; when they have become too warped, the card reader of the computer will no longer be able to process them. As a result, as soon as the study is put on cards, the cards are read into the computer, which transfers the data to a less bulky and more reliable storage medium, such as magnetic tape.

Magnetic tapes and disks are the most reliable storage media. The tapes are very similar to those used on tape recorders, except that computer tape is twice as wide and is on a larger reel. The 67,500 cards in the example above could be put on about 324 feet of tape, less than one-seventh of a standard 2,400-foot tape reel. Other studies could also be put onto the same reel of tape. Magnetic disk packs look like a stack of phonograph records, but, like the tapes, store information magnetically.

DATA ANALYSIS

For those in awe of computers, we should stress that a computer is a helpless piece of machinery. For our purposes, it is essentially a high-speed calculating machine for processing numerical information. It is a very useful tool because it can be programmed to perform a variety of tasks.

Computer Programs

A program is an extremely detailed set of operations to be followed by the computer. For example, a relatively simple program can direct the computer to read in all of the survey data and then count how many people gave each response for each variable—such as how many people are coded "2" in column 70 of the twelfth card for each respondent. More complex programs can do all of the types of analysis presented in this book.

Most computer installations have a set of statistical programs that can be used to perform common functions, such as those described in later chapters. This means that it is usually not necessary for social scientists to write their own programs in FORTRAN or some other computer language. They need only learn to use (and write the "control cards" for) the programs that are already available. But, as we shall see, having the program is only the beginning of the analysis.

In using these programs, we must sometimes remind ourselves that computers lack intelligence. The computer must be told what to do—and in terms it can interpret. A computer might perform some statistical task if a "1" were punched in column 50 of a control card or if a particular word were punched on a control card. However, if there is the slightest punching error, the computer will not perform the desired task and may terminate the job. Computers do what they are told, precisely what they are told, and nothing more. They cannot figure out what we wish we had told them to do.

Computers also cannot determine what statistical tests or operations are appropriate for our data. We must decide ourselves on the type of analysis. The program we use will probably be capable of printing out a large number of statistical results, but we must decide which of the statistical results are appropriate given the limitations of our data. For example, if our data represent a sample from a larger population, we might want to estimate the probability of getting our results by random chance. If our results were so weak that they could have occurred by chance, then they would be of no significance. Computer programs often print out tests of significance, but most tests of significance that are written into computer programs are only appropriate to simple random samples (which are the exception rather than the rule). Similarly, if we have data for the entire population of interest (so there is no sampling-error problem), we would probably choose to ignore tests of significance, even if the computer calculated them.

To reiterate, computers do not think or analyze or make decisions. Computers can only do the simple tasks of counting and calculation.

People must do the thinking about their data and analysis. What to calculate and what the answer means are still the work of the researcher, not the equipment. Keep in mind that those who defend their analysis solely on the grounds that it was "done on a computer" probably do not know or understand what the computer did.

Computer Operating Modes

Most computers operate in a *batch mode*. This means that various users (1) submit a set of punched instruction cards to the computer operator, requesting certain analysis, and then (2) return later for the printed results. Processing the job may take a few minutes or hours, depending on how heavily the computer is being used. Rarely does social-science analysis use more than a couple of minutes of computer time.

Many computers now also operate in an *interactive mode,* which permits direct communication with the computer. The user can sit at a terminal, connected by telephone to the computer, type instructions directly to the computer, and get results back immediately. Some of these terminals look like large electric typewriters; some are teletype machines like the ones the wire services (AP and UPI) use; others resemble television sets with typewriter keyboards attached. This latter type of terminal does not produce output on paper but writes instead on the television-type screen. The terminals are connected to the computer through normal phone lines. The user simply "dials in" to the computer and then connects the terminal to the phone.

The interactive mode of the computer permits conversational control of programs. The computer can be programmed to use English phrases to ask the user for clarification of an instruction that it could not interpret. Similarly it can ask the user for the next thing it is to do. The most sophisticated programs might directly ask the user, "Which variables do you want to correlate?" and then implement the correct programs if the user replied, "Party identification and vote."

Computer Instructions

Although we cannot give precise instructions about how to use a specific computer, we can outline the basic elements of what all computers must be told in order to get them to work for you. The first distinction that is useful is between instructions to the computer's operating system and instructions to the particular statistical program being used. (Unfortunately, both sets of instructions are commonly—and confusingly—referred to as "control cards.") Table 4.1 shows instructions to the operating system and to the statistical program.

Operating-System Instructions. The first instruction to almost all computer operating systems is a signon ("sign on"), logon ("log on"), job, or run instruction. The particular name is not important. The purpose of this instruction is to identify the user as a legitimate computer user and to give the account number that should be charged for the job. This instruction also starts the job.

Next, a set of instructions tells the operating system where the data are and the name of the statistical program that will do the analysis. (Some systems need the data location first; others need the program name first.) After processing both pieces of information, the operating system starts the execution of the statistical program. When the statistical program is finished, it returns control to the system. The last system instruction is to end the job. At this point, the computer calculates the charges for the run, bills the user's account, and then goes on to the next job submitted by some other user.

Program Instructions. The instructions to the statistical program usually start with a title. This instructs the program to print

Table 4.1 Organization of Jobs for Computer

Computer's Operating System

(1) Signon, Run, Logon, or Job card
- starts the job
- gives an account to charge for the run
- identifies user

(2) Program and data instructions
- names the desired program
- gives the necessary tape (or other file) information
- starts execution of the statistical program:

Statistical Program
(*a*) Title
(*b*) Operations—tables, marginals
(*c*) Variable list
(*d*) Options and statistics
(*e*) Finish and return to system

(3) End of job card—done using the computer

the identifying title at the top of each page. Another instruction tells the program what type of analysis is to be done—whether it is to construct tables, calculate marginal frequencies (defined in later chapters), or what. The next instruction contains the list of variables that are to be used in the analysis. Then the program must be told which options are desired with the analysis. For example, a program might graph some of the data if a specific option were requested. Although some programs automatically print all the possible statistics, others permit the user to select some and omit others. For example, some table-generation programs must be told which statistics (such as tau-*b*, gamma, and lambda as described in Chapter 7) are to be calculated. The last instruction to the statistical program is to return control to the computer's operating system. The above is just an outline of what many statistical-analysis programs require; some programs need additional instructions and others fewer. Also, the order of the instructions vary somewhat from program to program.

Computer Use

Required Information. To use the computer, a researcher needs (1) instructions on how to use the local computer (that is, operating-system instructions), (2) a "write-up" or manual describing the instructions for the statistical program(s) to be used, and (3) a "codebook" for the data that describes the variables and their categories. There are many different computer manufacturers (IBM, Control Data, Univac, Honeywell, Digital Equipment Corporation, and so on), and each company has different internal specifications and different instructions for its machines. IBM has a standard "operating system" for its 360 and 370 series computers, but some installations have their own operating systems. Because of the variation from system to system, we cannot give more detail on precisely what operating-system instructions are needed for the computer that you might use.

Similarly, there are no standard statistical programs. A program that operates on an IBM computer may or may not work on a Control Data (CDC) computer. Even if a program does work on more than one type of machine, some of the specific instructions may be different for different computers. A few large-program packages—BMD (Bio-Medical), OSIRIS, and SPSS (Statistical Package for the Social Sciences)—are now available on many computers, but there is no universal social-science program, and each program has its own operating instructions. Although the basic outlines of instructions are similar, the syntax of the instructions will be based upon decisions made by the author of the program as to which choices should be available to the users and how the selections should be made.

The final information required by the computer—the specific variables to be analyzed—depends on the study being analyzed. Published codebooks are available for major studies. They indicate the available variables and their categories. Looking at the codebook for the 1972 election study, for example, we see that presidential vote is variable "478." An analysis of that variable might refer to presidential vote as "V478," depending on how the data tape is set up and how the analysis program is written.

An Example. Let us consider the SPSS program package for finding the number (frequency) of people who have given each response for a question. SPSS is one of the most widely available and easy-to-use programs. This example shown below would produce the frequencies for presidential vote (V478) and other variables (V490–V493) for the previously saved file ELEC72.

```
RUN NAME         VOTE AND OTHER VARIABLES
GET FILE         ELEC72
FREQUENCIES      GENERAL = V478, V490 TO V493
OPTIONS          8
STATISTICS       1
FINISH
```

The first card (RUN NAME) is the title card described above. The second (GET FILE) tells the program the name of the data file. The third (FREQUENCIES) indicates the type of analysis that is to be done *and* the variables to be analyzed. The next (OPTIONS) requests option 8, which in this case is a graph for each variable. The STATISTICS card requests that statistic 1, averages, be calculated for each variable. The FINISH card completes the SPSS run. These setup cards are slightly different from the outline in Table 4.1—which illustrates the lack of uniformity in computer programs.[3]

[3]Unfortunately, we are unable to illustrate system instruction cards. They vary so much between universities that any example would be wrong at most universities and thus would confuse matters more than it would help.

SUMMARY

We did not intend to present enough material in this chapter to enable the reader to walk into a computer center and operate a computer, but we did intend to present a general view of how computers fit into the analysis process. By studying the detailed instructions for a local computer system and its analysis programs, the reader should be able to learn to use the computer. Most social scientists use the computer extensively in large-scale analysis, and it is a tool worth understanding. However, as we have stressed, it is only a tool. Computer use is largely mechanical, but the design of an analysis that explores the causal underpinnings of attitudes or behavior is a sophisticated and substantive problem for the analyst.

Questions

1. Get a computer card and punch on it the data shown in Figure 4.1. (Try this and you will see that computer use is much easier than you might have thought.) Punch your name on another card to see how letters are handled.
2. Find out what your local computer's operating instructions would be for running SPSS (or whatever program system is most heavily used for social-science analysis at your installation). If you can get access to some computer time, try a simple run and see what happens. (Our experience to date shows that computers and first-time computer users survive the experience fairly well.)

5

Single-Variable Statistics

The simplest kind of analysis that can be performed is to look at one variable at a time. This could consist of tallying up the responses on a particular question on a survey or calculating the mean GNP for postwar France. Clearly this type of analysis is *not* the main part of data analysis but is usually a first step. Unfortunately, for some social scientists it is the last step in their analysis. However, most social scientists are interested, at a minimum, in the relationships among the variables they are studying.

FREQUENCY DISTRIBUTIONS

As a way of introducing single-variable statistics, we shall begin with an example from survey research. The simplest display of the results for a single variable is a list showing the number of people giving each answer to the variable—the distribution of the frequency of each response. This information is sometimes called the "marginals" because the frequency distribution is often printed in the margin of the codebook. Table 5.1 consists of some hypothetical data that will be used to illustrate several points about frequency distributions.

The table shows that a majority of people interviewed supported the idea of government health insurance. Of 1,303 people, 863 supported government health insurance; 863 is clearly bigger than 440, the

Table 5.1 Attitudes on Government Health Insurance

"Would you support or oppose having the federal government take over health insurance in this country?"	
Support	863
Oppose	440
	1,303

SOURCE: Hypothetical.

number that opposed government health insurance. This summarizes the results nicely, except that it takes a while to decide whether or not 863 is a very large majority of 1,303.

Percentage Distributions

It is usually more effective to express the results as percentages and present the frequency distribution in percentage form. For example, the proportion supporting the program in Table 5.1 is $(863 \div 1{,}303) \times 100 = 66.2$ percent. The proportion opposing it is $(440 \div 1{,}303) \times 100 = 33.8$ percent. Table 5.2 again reports exactly the same data as does Table 5.1; however, this time the display is in percentages.

The second table shows at a glance that government health insurance commands approximately a two-thirds majority. The number of cases (often called the N) is listed under the percentages so that the reader knows immediately how much confidence to place in the results. A two-thirds majority is a large majority, but we have much greater confidence in the results when we know that there were 1,303 respondents rather than just 30 or 40. Listing the number of cases also allows the reader to reconstruct the actual numbers upon which the percentages are based.

Table 5.2 Attitudes on Government Health Insurance

"Would you support or oppose having the federal government take over health insurance in this country?"	
Support	66.2%
Oppose	33.8%
Total	100.0%
(Number of cases)	(1,303)

SOURCE: Hypothetical.

Missing Data

Unfortunately, survey results are never as cut and dried as those of Tables 5.1 and 5.2. People may say that they would support the program if private insurance was still permitted. A "depends" category might be created to account for them. Many people have not thought about the problem, and some of them will admit that they have "no opinion." Sometimes an interviewer forgets to read a question, or a respondent terminates the interview before the interviewer has managed to ask this particular question—if so, the respondent's attitudes on this item are "not ascertained."

If we were to include these categories in the frequency distribution, we might obtain the results shown in Table 5.3. How do we summarize opinion in this table? The largest problem is that people in the "no opinion" and "not ascertained" categories have not given their opinions on the question.

The "Not Ascertained" Category. The "not ascertained" group is particularly uninteresting, since all it shows is that the question was not asked. Therefore, that group could certainly be treated as "missing data" and dropped from the table. Replacing the remaining numbers with percentages would give us Table 5.4. Note that the number of cases on which Table 5.4 is based is less than the number on which Table 5.3 is based because the "not ascertained" category has been subtracted.

Incidentally, why do the percentages in Table 5.4 add to 100.1 percent rather than 100.0 percent? Because we have rounded each percentage to one digit after the decimal point, and the rounded percentages sum to 100.1. It is common in such tables for percentages to

Table 5.3 Attitudes on Government Health Insurance

"Would you support or oppose having the federal government take over health insurance in this country?"	
Support	863
Depends	223
Oppose	440
No opinion	350
Not ascertained	18
Total	1,894

SOURCE: Hypothetical.

Table 5.4 Attitudes on Government Health Insurance

"Would you support or oppose having the federal government take over health insurance in this country?"

Support	46.0%
Depends	11.9%
Oppose	23.5%
No opinion	18.7%
Total	100.1%
(Number of cases)	(1,876)

SOURCE: Hypothetical.

sum to 99.9 percent or 100.1 percent instead of exactly 100.0 percent because of the effects of rounding off to the nearest tenth of a percent. Similarly, if the percentages are rounded to the nearest whole percent, the totals might be 99 percent or 101 percent. This error due to rounding should not be of any concern.

Table 5.4 gives a very honest rendition of the results. More people support government health insurance than oppose it. But many have intermediate views, and many have not made up their minds yet. Those supporting the program are not a majority in themselves, but could form a majority by attracting some of those in the "no opinion" or "depends" categories to their point of view.

The "No Opinion" Category. The "no opinion" category shows to what extent opinion has crystallized on the question. It is often critically important to show that the public has not made up its mind on an issue. However, we still may be interested in the distribution of opinion among those who have made up their minds. Of those who had an opinion and answered the question, how many support government

Table 5.5 Attitudes on Government Health Insurance

"Would you support or oppose having the federal government take over health insurance in this country?"

Support	56.6%
Depends	14.6%
Oppose	28.8%
Total	100.0%
(Number of cases)	(1,526)

SOURCE: Hypothetical.

health insurance? This means treating the "no opinion" category as a "missing data" category and excluding it from the table. Thus we now get Table 5.5.

Of those with opinions, a majority support the government health insurance plan and less than 30 percent oppose it. Table 5.5 does not tell how many people are undecided on this issue, but it does summarize the views that were stated. One would present a table without the "no opinion" category when the assumption is that the opinions of the "undecideds" will eventually be distributed in about the same way as the opinions of those who now have opinions. In other words, a table like 5.5 is based on the assumption that about 56.6 percent of the people with no opinion will eventually support government health insurance. Although that is the most likely assumption, there are several other assumptions that also prompt one not to report the "no opinion" category. Imagine a policy maker who takes a survey to help her come to a decision which must be made soon. She can safely ignore those respondents who have not made up their minds yet and those who do not care enough to have an opinion. Another example is provided by a survey taken the day before an election. It is doubtful that those who have no opinion about the candidates will take part in the election.

There are also certain circumstances under which the "no opinion" category definitely should not be omitted. Dropping the "no opinion" respondents would be misleading if there were reason to suspect that they would eventually have a particular opinion. For example, one might not want to drop the "no opinions" on a question concerning racial attitudes if there were reason to believe that "no opinion" was just a covert racist answer—the respondent claimed to have no opinion rather than give the interviewer a racist answer. This is rarely a serious problem, although it often makes sense to consider whether "no opinion" has such a concealed meaning. Also, "don't know" may be a meaningful answer, as on political-information questions—such as how many justices there are on the United States Supreme Court. The answer "don't know" is not missing data but rather an indication of the fact that the respondent did not know the correct answer. It would thus be a mistake to drop that category from the analysis. Another circumstance in which one should not drop the missing-data category is when it is quite large in comparison with the substantive categories. That usually indicates some problem with the question, and the reader should be made aware of the potential problem.

Note that Tables 5.1 through 5.5 are all based on exactly the same set of responses, those listed in full in Table 5.3. The tables differ in only two respects: (1) whether the results have been expressed as percentages to simplify analysis and (2) which categories have been omitted. Table 5.2, 5.4, or 5.5 could be published as "the distribution of public opinion on government health insurance," yet the three tables do give somewhat different views of the nature of public opinion. When you read a poll result in the paper or elsewhere, it is worth checking what they did with the intermediate categories (like "depends") and the missing-data categories.

As a final means of displaying the same data, Table 5.6 gives the distribution of actual opinions plus an indication of how crystallized opinion is. This may be the most honest and helpful way to display the results.

Interpreting Frequency Distributions

We must not overinterpret the percentages in such tables. They are affected by many potential problems, two of which we should emphasize here. The first is sampling error, the error that occurs when trying to describe a population with a sample. It depends mainly on the sample procedures and sample size. For example, Table 5.6 says that 56.6 percent of those with opinions supported government health insurance. But with 1,526 interviews, there is a sampling error of approximately 3–4 percent, which means that the actual percentage

Table 5.6 Attitudes on Government Health Insurance

"Would you support or oppose having the federal government take over health insurance in this country?"

Support	56.6%
Depends	14.6%
Oppose	28.8%
Total	100.0%
(Number of cases)	(1,526)
(No opinion)	(18.7%)
(Total number of cases)	(1,876)

SOURCE: Hypothetical.

in the population supporting government health insurance is between about 53 percent and 60 percent. The "depends" category is between 12 percent and 18 percent, and the "oppose" category is between 26 percent and 32 percent. Regardless of the sampling error, it is clear that the program has majority support (of those with opinions) and that the opposition is below the one-third level. That kind of information is usually enough. Rarely do we require more exact estimates.

Also, we should not be overly concerned with the exact results, because they depend a great deal upon the exact wording of the question (or code categories). What if the question on health insurance had been: "Would you prefer government health insurance or private health insurance in this country?" Perhaps this wording makes the alternative to government health insurance clearer than "Would you support or oppose having the federal government take over health insurance in this country?" If all respondents had carefully researched health insurance programs, the two wordings might yield identical results. However, if you assume that few people have given the question much thought before the interviewer asks them, it is not at all unreasonable to expect a 10–20 percent difference due solely to question wording. (There may be no difference at all in the results

Table 5.7 1972 Reported Vote for President

Category	Number
Nixon	1,021
McGovern	566
Schmitz; Wallace	11
Democratic candidate	0
Republican candidate	0
Refused to say	19
Other candidates	7
Don't know	1
Not ascertained (includes those who voted but not for President)	37
Inappropriate	
Did not vote	623
Not reinterviewed after the election	420
Total	2,705

NOTE: The inappropriate category was separated into the two groups shown here on the basis of another survey variable.
SOURCE: Center for Political Studies, 1972 American National Election Study.

Table 5.8 1972 Reported Vote for President

Nixon	64.3%
McGovern	35.7%
Total	100.0%
(Number of cases)	(1,587)
(Not voting)	(28.2%)
(Total number of cases)	(2,210)

SOURCE: Center for Political Studies, 1972 American National Election Study.

of two different wordings, but we should not count on it.) Thus, Table 5.6 should be interpreted as showing public support for government health insurance, but we should not give too much emphasis to the 56.6 percent level of that support.

Another Example

Let us turn to another example to help clarify the techniques of displaying frequencies. Table 5.7 gives the frequencies for the responses to the presidential-vote variable from the 1972 CPS Election Study. These results can be arranged in a more concise manner than in Table 5.7. We might present only the percentages of people voting for each of the two major party candidates. (If the American Independent Party candidate had received more votes, we probably would have wanted him included in the percentages. As it is, he received less than one percent of the vote.) Since a large number of the people did not vote, it is important to include that fact in some way. In Table 5.8, percentages have been calculated, and irrelevant categories like "don't know" have been eliminated.

Obviously there are other reasonable ways of presenting these data; however, Table 5.8 conveys the points that would interest most readers: (1) the percentage voting for each of the two major candidates and (2) the percentage not voting. Note that these percentages would have been slightly different if we had included the vote for the AIP candidate. (It is also worth noting that the above data are the result of asking people whether they had voted and for whom. A check of poll lists nationwide would indicate that a substantially higher percentage of people did not vote than is reported here. A portion of that difference is a result of people feeling as though they should have voted and thus reporting having done so to the interviewer.)

Although the examples in this section have been from survey research, the problems are in no way unique to that field. For example, a social scientist interested in comparative politics might well want to see the frequency distribution on a variable such as type of national government and would face the same kind of problems faced by the survey researcher. How do you treat the missing data? There might even be two or more types of missing data: (1) countries that could not be classified as to type and (2) countries that were inadvertently omitted. How much do the results depend on the specific coding details? In any case, the problems are structurally quite similar no matter what the source of the data or the substantive area.

In general, when constructing frequency distributions, it is most important to include sufficient information to illustrate the important substantive point. Often what you want to show with the data will determine how much detail (and, in fact, which details) should be included in a table. Readability is usually improved by showing percentages rather than raw frequencies, although the number of cases underlying the percentages should be given, so that the reader can reconstruct the raw frequencies if necessary.

LEVELS OF MEASUREMENT

It is easy to examine Table 5.8 (or any preceding tables in this chapter) and get a quick impression of the distribution of public attitudes. Consider, however, Table 5.9. Here, there is a relatively large number of substantive categories. One can examine the full set of percentages, but it would be useful to summarize the results more compactly. In this example or in any situation in which we have added up the responses and produced a frequency distribution, we may want to summarize those responses statistically. We might want to report the average response, or give some idea of how unified the public is in its attitudes. This gets us into the realm of statistics, and it also focuses our attention on the limitations of the specific questions that we asked.

The first problem in an analysis is to decide upon the appropriate statistical operations. Procedures appropriate to the analysis of some

Table 5.9 Ideal Family Size

"How many children would you consider the right number for a family?"

None	10.5%
One	22.1%
Two	35.2%
Three	21.6%
Four	5.8%
Five	3.1%
Six	1.6%
Eight	.1%
Total	100.0%
(Number of cases)	(1,404)

SOURCE: Hypothetical.

variables are inappropriate to the analysis of others. For example, we could calculate from Table 5.9 the average number of children mentioned as the ideal, but we could never calculate the "average religion" of the respondents. Though the average may be a useful concept in analyzing some variables, it cannot or should not be calculated for others. What measures are appropriate depends mainly on the type of measurement employed for the variable in question. Statisticians distinguish between at least three basic "levels of measurement": (1) interval, (2) nominal, and (3) ordinal. Each level requires a different type of statistical analysis.

Three Measurement Levels

Variables that are intrinsically numerical, like age and income, are known as *interval variables.* Interval measurement occurs when the responses are represented by actual numbers and the distance between the successive numbers is equal. Such numbers can be manipulated arithmetically by addition and subtraction; averages can be calculated, and even more complex statistical techniques can be used. The problem, however, is that very few of the variables that social scientists deal with are interval variables.

By contrast, categorical variables such as religion are known as *nominal variables.* A person may be Catholic, Protestant, Jewish, a member of another religion, or a member of no organized religion; there is nothing at all numerical about these categories, even though a coder may have numbered the categories in order to distinguish one from another. In other words, there is no "average religion." Region is another example of a nominal variable. One may live in the North, East, South, or West, but there is nothing numeric about these categories. Even if such categories have been numbered in the coding process, arithmetic operations like averaging are not appropriate.

The third important type of measurement occurs when people are sorted into ordered categories that do not have any intrinsic numerical qualities. People might be asked if they "strongly agree," "agree, but not strongly," are "neutral," "disagree, but not strongly," or "disagree strongly" with a statement. Such *ordinal variables* are categories with intrinsic order but without intrinsic numerical properties because there is no way to ascertain that the differences between successive categories are the same. For example, the difference between strong and weak agreement with a statement may be more or less than the difference between weak agreement and neutrality. Ordinal measurement is said to be stronger than nominal measurement, since there is at least an order to the categories; but it is weaker than interval measurement, since the categories are not meaningfully numbered.

The distinction between nominal, ordinal, and interval measurement reappears continually in statistics, so it is important to see the differences as clearly as possible. Consider three possible questions regarding religion:

a. What is your religion? Are you Protestant, Catholic, Jewish, a member of another religion, or a member of no religion?
b. How religious do you consider yourself? Do you consider yourself very religious, somewhat religious, or not at all religious?
c. How many times a month do you usually go to religious services?

The first question is aimed at classifying the person in terms of nominal religion categories. The second question seeks to determine the person's religiosity in terms of self-classification along an ordinal continuum. The third question measures attendance at religious services in direct interval terms. Remember that because a coder has assigned Protestants to category 1, Catholics to 2, Jews to 3, others to 4, and no religion to 5 does not mean that religion is an interval variable. The numbers in this case are just short-hand names for the categories. We could just as well call the Catholics group B as group 2. In order for the level of measurement to be interval, the numbers must have intrinsic meaning and not just be assigned arbitrarily.

Choosing a Measurement Level

The three levels of measurement represent a hierarchical set of rules according to which the highest level, interval data, can be divided into ordered categories and treated as ordinal data, a lower level. Similarly, ordinal data can be treated as nominal data, the lowest level. Shifts in the opposite direction are generally not possible, however.

The three levels of measurement were introduced by saying that different types of statistics are appropriate for different levels of measurement; throughout the remainder of this book, we shall refer to levels of measurement in conjunction with the question of which statistical procedures are appropriate. How strictly must these hierarchical rules be followed? Some researchers apply them with zeal; others relax them under certain circumstances. In particular, for reasons reviewed at the beginning of Chapter 11, some analysts frequently apply interval statistics to ordinal variables. Although we do not necessarily disagree with that practice, it is fairly sophisticated, and we believe for pedagogical reasons that the most strict approach should be mastered first.

Finally, the level of measurement does not matter for *dichotomous variables*—those with only two meaningful categories. Sex, major party vote (Democrat or Republican), and whether or not a person voted are three examples of such variables. They can be considered interval, ordinal, or nominal in many instances without any loss of meaning. If, for example, you code male as "0" and female as "1," then an "average sex of 0.55" would be equivalent to 55 percent of the sample being female. Calculating an average for this dichotomous variable would make sense, so long as you do not think of any person as being the average sex.

MEASURES OF THE CENTRAL TENDENCY

The simplest type of summary measure is a measure of the "central tendency"—one that indicates how the typical person behaves or indicates the typical value of the variable. The most familiar measure of central tendency is the arithmetic average, known technically as the *mean*.

The Mean

The mean can be calculated only for interval variables. To compute it, you add up the values of the variable for each case and divide that by the number of cases. Thus, if a survey of six people found that they attended religious services 0, 1, 3, 3, 5, and 8 times per month, you would add these numbers together to get a sum of 20 and divide by 6 (the number of people) to get a mean of 3.33. The average number of times that these people went to church is 3.33. (We leave it to the reader to consider the philosophical implications of going to church .33 times.) Table 5.10A illustrates the above calculation.

Similarly, in Table 5.9, we saw that we needed a summary measure or at least a more concise way of describing the information in the

Table 5.10 Central Tendency Measures

(A) Interval Variable: Frequency of Attendance of Religious Services

	Times per Month	Frequency	(Times × Frequency)
x_1	0	1	$(0 \times 1) = 0$
x_2	1	1	$(1 \times 1) = 1$
x_3, x_4	3	2	$(3 \times 2) = 6$
x_5	5	1	$(5 \times 1) = 5$
x_6	8	1	$(8 \times 1) = 8$
Total		6	20

$$\text{Mean} = \overline{X} = \frac{\sum_{i=1}^{N} x_i}{N} = \frac{20}{6} = 3.33$$

(B) Ordinal Variable: Religiosity

	Proportion	Cumulative Proportion	
Very religious	40%	40%	
Somewhat religious	35%	75%	(50% point)
Not at all religious	25%	100%	

Median: Somewhat religious

(C) Nominal Variable: Religion

	Proportion
Protestant	45%
Catholic	30%
Jewish	5%
Other	10%
None	10%

Mode: Protestant

SOURCE: Hypothetical.

table. The mean should help. Although the mean for Table 5.9 is a little more time-consuming to calculate than the one in Table 5.10A, the process is exactly the same. After adding up the number of children mentioned by each of the 1,404 respondents and dividing that number by 1,404, a mean of 2.06 was determined.

A comment on notation is called for here. Although the measure that we are discussing here is not very complex, it is useful to understand some of the mathematical notation that is often used to describe the mean. If the notation can be mastered for this simple measure, then it will be much easier to understand the notation for some of the more complex ones. We often speak of the observations of the variable X as x_1 (the first person's value), x_2 (the second person's value), $x_3, x_4, \ldots, x_N$. We would say that there are N observations of X, where N is the number of people. The sum of those N observations of X can be written:

$$x_1 + x_2 + x_3 + x_4 + \ldots + x_N = \sum_{i=1}^{N} x_i$$

This is read as follows: "the sum of the x sub i's, where i goes from 1 to N." The mean of X, often written $\overline{X}$, can be expressed:

$$\overline{X} = \frac{\sum_{i=1}^{N} x_i}{N}$$

Although the notation is different, the result is still the same as what is commonly called the *average*, or the *mean*.

Alternative Measures

Although the average is useful for interval-level data, other measures are available for ordinal and nominal data. If the variable is ordinal, we can look for the *median*—the middle position. Thus if 40 percent of the sample indicate that they consider themselves very religious, 35 percent somewhat religious, and 25 percent not at all religious, then the middle person in the sample considered himself somewhat religious. That is the median response. (See Table 5.10B.) Technically, the median response should have half of the remaining responses below it and half above it. What frequently happens, however, is that, as in our example, the median falls in a large group of other responses, so that the numbers of responses above and below are not exactly equal. In this instance, what we have found is the category that comes closest to being in the middle of the distribution.

If the variable is nominal, all we can do is to find the *mode*—the category that occurs most frequently. If more people in the United States consider themselves Protestants than Catholics, Jews, members of other religions, or members of no religion, then the modal religion is Protestant. (See Table 5.10C.)

Choice of Central Tendency Measure

Sometimes the mode and median are even used for interval-level variables. For example, Table 5.9 on ideal family size might best be summarized by saying that both the modal number and the median number of children desired is two. The mean number of children desired in this case is also quite close to two—it is 2.06. In some instances, the mean can be unduly influenced by a few extreme values of the variable that are outside the normal range of values. Say, for example, that a few people said that twenty children would be ideal; that might raise the mean so high that the mode or median would be a much more reasonable measure of the view of the average person.

When social scientists use statistics, they are often choosing among measures already calculated by a computer rather than deciding which ones should be calculated. This is because people who write computer programs often decide to have the computer automatically print a wide range of statistics and let the researcher decide which to use. Therefore, we must not assume that we have a true interval-level measurement on a variable just because the computer calculated a mean for it. The computer can calculate a mean for *any* set of numbers—whether they are interval variable values or just category names. The fact that a measure was calculated does not give it substantive justification—it may be nonsense.

Missing Data

In computing any of these measures of central tendency, we must always be sure that the missing data have been excluded. For example, a set of data might have the symbol "99" coded for a person who was accidentally not asked a question ("not ascertained"). That "99" should not be allowed to affect the calculation of, say, the average number of children a person wants!

MEASURES OF DISPERSION

Social scientists are interested in seeking to account for the differences among people in their answers. The *variance* is a measure of how different the scores are for interval variables. Not everyone has the same score, so there is variance. If everyone's score was the same, that score would be the mean, and there would be no variation. Consequently, the variance is a measure of how dispersed the cases are from the mean. Since the mean is a measure of the central tendency, it is near the middle of the cases; the variance tells us how scattered the cases are around the mean. The smaller the variance, the closer the cases are to the mean; the larger the variance, the more widely they are scattered.

We might try to measure the dispersion around the mean by subtracting the mean from each value and summing those differences. However, we would find that the sum would be zero (within rounding error) because the sum of the differences for the cases above the mean is the same as the sum of the differences for those cases below the mean (with the opposite sign). This can be seen in the third column of Table 5.11. There are at least two ways of handling this problem. One is to take the absolute values of the differences, and the other is to square them. Although it would make little difference

Table 5.11 Variance of Age

Age	Frequency	(Age − Mean)	$(\text{Age} - \text{Mean})^2$
20	1	$(20 - 24.33) = -4.33$	$-4.33^2 = 18.75$
23	1	$(23 - 24.33) = -1.33$	$-1.33^2 = 1.77$
30	1	$(30 - 24.33) = 5.67$	$5.67^2 = 32.15$
Total	3	0.01	52.67

$$\overline{X} = 24.33 \qquad s^2 = \frac{\sum_{i=1}^{N} (x_i - \overline{X})^2}{N} = \frac{52.67}{3} = 17.56$$

SOURCE: Hypothetical.

Table 5.12 Income of Two Groups

Group A	Group B
\$9,800	\$5,000
\$9,900	\$7,500
\$10,000	\$10,000
\$10,100	\$12,500
\$10,200	\$15,000
$\overline{X}$ = \$10,000	$\overline{X}$ = \$10,000
s^2 = 20,000	s^2 = 12,500,000
s = 141	s = 3,535

SOURCE: Hypothetical.

for present purposes, statisticians prefer working with the squared values when they generalize the variance concept to more than one variable. Therefore, the variance is customarily defined as the average squared deviation from the mean:[1]

$$s^2 = \frac{(x_1 - \overline{X})^2 + (x_2 - \overline{X})^2 + \ldots + (x_N - \overline{X})^2}{N} = \frac{\sum_{i=1}^{N} (x_i - \overline{X})^2}{N}$$

Table 5.11 illustrates this calculation. The mean is subtracted from each person's score; the resulting differences are squared, summed, and then divided by the number of people. For the first person, the mean of 24.33 is subtracted from the score of 20. That yields a deviation of −4.33, which, when squared, is 18.75. Similar calculations for the second and third persons give squared deviations of 1.77 and 32.15. These squared deviations sum to 52.67. The variance is 52.67 divided by 3 (the number of people), which is 17.56.

The example in Table 5.12 illustrates two sets of data on income with the same number of people (5) in each and the same mean income (\$10,000) but with quite different variances. Group A has a variance of 20,000 while group B's variance is 12,500,000. Both of these turn out to be large numbers, although it is still clear that the variance for group B is far larger than for group A. Since we squared the differences between each value and the mean before we added them, statisticians would suggest that it would be more meaningful

[1]Note that the denominator of the variance in this equation is N because this is the variance for a population. Some versions of the formula have $N-1$ in the denominator; technically, those refer to samples. Usually, the sample size is so large that the numerical difference between the two formulae can be ignored. When we are specifically discussing a sample, we will use the $N-1$ version; otherwise, we will use N.

to look at the square root of this variance. The resulting statistic is called the *standard deviation.* The standard deviations for the two groups are 141 and 3,535 respectively. Some feel that these values make more intuitive sense as measures of dispersion. Looking at these measures, the incomes for group A are obviously substantially less dispersed than for group B.

The variance and the standard deviation can be computed only for interval-level variables because the data must be actual numbers. Yet people's responses vary even when they are not an interval variable, so there is still something important to explain. For ordinal-level variables, we can seek to explain why one person's score is higher than another's. That is, for ordinal-level variables we seek to explain the order of responses. For nominal variables, we seek to explain why different groups have different modal values on the variable.

Until now we have been discussing situations in which a variable has some variance. However, we occasionally encounter variables with little or no variance; that is, the responses are all nearly the same. Since our purpose is to explain variation, there is little to explain in those instances, and similarly, those variables have little explanatory power. Such variables are not useful in data analysis. For example, there is no reason to try to explain the vote for president of Republicans in 1972, because nearly 94 percent of the Republicans voted for Nixon.

STATISTICAL INFERENCE FOR MEANS

Recall that when estimating a proportion for a population from a sample there is a certain amount of inevitable sampling error, so all that one can say with confidence is that a proportion is within some range. It should not be a surprise that a similar situation holds for means. We shall not go into much detail, but we shall sketch the computation of sampling error when it is appropriate to compute means.

According to statistical theory, if you take a large number of samples and examine the means of a variable in the different samples, you will find that they follow the "normal curve" distribution of Figure 5.1.[2] In other words, most sample means will fall near the population

[2]More precisely, when the variance is unknown, the means follow the t-distribution, which is approximated well by the normal distribution for samples of more than 120 respondents.

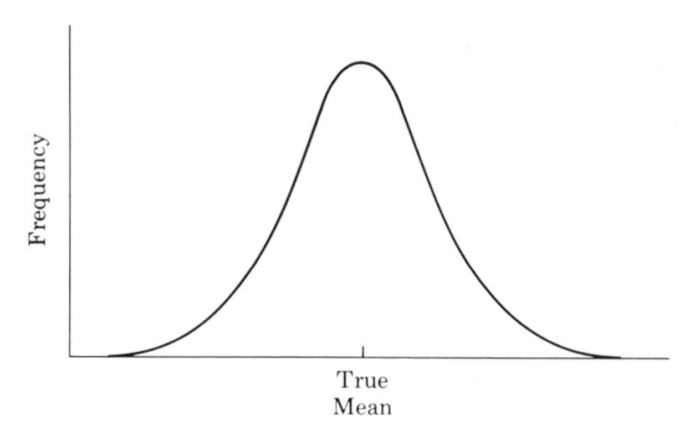

Figure 5.1
The Distribution of Sample Means (A Normal Curve)

mean; few will be far from the population mean. The "standard error of the mean" gives the dispersion of the sample means:

$$s_m = \sqrt{s^2/(N-1)}$$

For large samples, it can be shown that 95 percent of the sample means will fall within 1.96 standard-error units above or below the population mean. Using reverse logic, we can estimate a *confidence interval;* there is a 95 percent chance that this confidence interval includes the true population mean.[3] This confidence interval is centered around the sample mean and extends 1.96 standard-error units, above it and below it. Thus, if a sample mean is 25.0 with a standard deviation of 10.0 for a sample of size 401, the standard error is 0.5 and there is a 95 percent chance that the population mean is between [$25 - (1.96 \times 0.5) =$] 24.02 and [$25 + (1.96 \times 0.5) =$] 25.98. Note that the width of the confidence interval depends on the size of the standard deviation and the size of the sample. A smaller standard deviation and a larger sample size will both work to permit more precise estimates of the population mean.

It should be realized that the statistical-inference procedures given here and later are based on simple random samples. Most large surveys are instead cluster samples or multi-stage area samples, so the exact significance procedures given here (and by most conventional com-

[3]This statement of the confidence interval is not in strict accord with classical statistics, but the practical implications of the precise difference are small.

puter programs) would not be applicable. In particular, the standard error would be larger and the 95 percent confidence interval would be correspondingly wider if the sample is not a simple random sample.

SUMMARY

The hierarchy of levels of measurement provides a valuable set of guidelines in the use of statistics. The appropriate statistics depend upon what has been measured and how. Ignoring the actual level of measurement entails the risk of error.

Questions

Partisanship	
Republican	233
Independent	306
Democrat	398

1. The total number of cases in this table is ____________.
2. What proportion of the people are Republicans? ____________.
3. What proportion are Democrats? ____________.
4. What proportion are Independents? ____________.
5. What categories have probably been omitted from this table? __________.
6. What is the partisanship of the average American according to these data? ____________.
7. Five people gave the incumbent president the following thermometer ratings: 95, 85, 80, 75, and 65. Find the mean thermometer score for the group, the variance, and the standard deviation.

6

Two-Variable Tables

Academic researchers are interested in understanding and explaining popular attitudes and behavior—not just describing them. Since attitudes and behavior are determined by the relationships among several variables, researchers must examine more than one variable at a time. This chapter begins our investigation of the relationships among variables with a discussion of the study of a pair of nominal or ordinal variables. The interval-variable case will be deferred to Chapter 11.

READING TABLES

A common way of studying two variables at a time is to look at a bivariate frequency distribution, which is also called a cross-tabulation or "cross-tab"—and more often is simply called a table. Cross-tabs are easy to read, but they are often misread. Table 6.1 was constructed from hypothetical data to illustrate how to read a table correctly.

Errors in Describing a Relationship

The table shows that more people were hurt crossing the street at the corner than were hurt jaywalking. In fact, 60 percent of those

Table 6.1 Pedestrian Accidents by Location of Pedestrian

	Crossing at Corner	Jaywalking	Total
Safe	1,997,000	198,000	2,195,000
Hurt	3,000	2,000	5,000
Total	2,000,000	200,000	2,200,000

SOURCE: Hypothetical.

hurt crossing the street crossed at the corner. Is it more dangerous to cross at the corner? A glance at the data would seem to suggest that it is. However, our intuition, all that we have read, and our mothers have told us that it is more dangerous to jaywalk—and so does the table if we read it correctly. Note that ten times as many people crossed at the corner (2,000,000) as jaywalked (200,000) and that only one-and-a-half times as many people were hurt at the corner. Our first glance missed the numbers of people who were crossing at each place.

A related misleading situation can arise when someone tries to demonstrate a relationship without presenting a table. For example, someone might argue that one should not cross the street at the corner because 60 percent of the accidents occur there. Without seeing the entire table, it is easy to accept inadvertently such a false conclusion.

Percentage Tables

Generally we find it easier to interpret a table if the data are expressed in percentages rather than in raw frequencies, like those in Table 6.1. Since we are interested in the relative dangers of crossing the street at the corner versus jaywalking, we would like to compare the safe passage of those who jaywalked with those who crossed at the crosswalk. Therefore, we should calculate the percentage of people who crossed at the corner who were hurt and the percentage who were safe. Next, we should calculate the percentage of people who jaywalked who were hurt and the percentage who were safe. Then we will be able to compare effectively the jaywalkers with those crossing in the crosswalk. In this way we should get a better idea of the relationship between the location at which one crosses the street and the likelihood of being hurt.

Table 6.2 Pedestrian Accidents by Location of Pedestrian

	Crossing at Corner	Jaywalking
Safe	99.85%	99.00%
Hurt	0.15%	1.00%
Total	100.00%	100.00%
(*N*)	(2,000,000)	(200,000)

SOURCE: Hypothetical.

In Table 6.2 it is clear that 1 percent of those who jaywalked were hurt, and only 0.15 percent of those who crossed at the corner were hurt. Crossing at the corner is almost seven times more safe than jaywalking. This hypothetical example illustrates some of the comparisons that can be made with data in table form. It is usually better to compare percentages than raw numbers; however, those percentages must be calculated in the right manner. For example, our previous statement that 60 percent of those hurt were crossing at the corner was correct but very misleading because it ignores the causal direction. The percentages should be calculated so that they add up to 100 percent for the categories of the causal variable—in this case, the location at which people crossed the street.

Social scientists use some specialized terms regarding tables partly in the hope that the terminology will be helpful in choosing the right direction for computing the percentages. Recall from Chapter 3 that the dependent variable is the one that we want to explain and that the independent variables are the ones used to explain the dependent variable. In the above example, the dependent variable being explained is whether one is hurt crossing the street; the independent variable is where the person crossed the street. Percentages were calculated within categories of the independent variable. Calculating percentages according to this rule permits the effects of the independent variable to be assessed properly.

COMPARING PERCENTAGES

We look at a table to determine how much effect one variable has on the other. This not only requires calculating the percentages in the right direction as just explained but also entails comparing those percentages. In this section, we explain how the extent of a relationship is gauged from a table. Our examples will be based on real data from the 1972 American Election Study conducted by the Center for Political Studies at the University of Michigan.

Percentage Differences

Table 6.3 reports 1972 data concerning voting turnout by sex. This table is meant to help us understand turnout rates—the proportion of the people who actually go to the polls and vote. That is the dependent variable. Our explanatory variable (or independent variable) in this instance is sex. It is possible that sex differences may cause turnout differences; so, sex is the independent (causal) variable. (It is difficult to imagine turnout causing sex differences.) The table shows that 76 percent of the men reported voting in 1972 compared to 70 percent of the women.

Percentage Direction. Note how the rule of the previous section was followed in percentagizing Table 6.3. We could have calculated the proportion of men voters and the proportion of women voters. Instead, we calculated percentages within the categories of the independent variable. Sex is the independent variable, one category of which is men. We calculated what percentage of the men voted and what percentage did not vote. We made similar calculations for women.

Table 6.3 Turnout by Sex 1972

(*A*) Column Percents

	Male	Female
Voted	76%	70%
Did not vote	24%	30%
Total	100%	100%
(Number of cases)	(975)	(1,308)

(*B*) Row Percents

	Voted	Did Not Vote	Total	Number of Cases
Male	76%	24%	100%	(975)
Female	70%	30%	100%	(1,308)

SOURCE: Center for Political Studies, 1972 American National Election Study.

Two forms of the cross-tabulation are shown in Table 6.3. Note that Table 6.3A has the independent variable as the column variable and that Table 6.3B has the independent variable as the row variable. Whether the data for the independent variable are in columns or in rows, the interpretation of the table is the same. Generally, we consider tables with the independent variable expressed by column headings easier to read (though the variable with the greater number of categories is often put on the rows to facilitate the printing of the table). Regardless of the choice, adopting a consistent table format within a paper or article is recommended.

Comparison Direction. Because we calculated the percentages within the categories of the independent variable, we can compare across categories to see the effect of the independent variable (sex) on the dependent variable (turnout). We know that 76 percent of the men voted and 70 percent of the women voted. We compare these two percentages to find a 6 percent difference, men voting more often than women by 6 percent. This is sometimes stated as the turnout rate for men being "6 points higher" than that for women. Another way to put it is that sex affected turnout by 6 percent in 1972. The 6 percent difference is, so to speak, the effect of sex on turnout.[1] We had to calculate the percentages in the correct direction in order for the percentage difference to show the effect of the independent variable on the dependent variable.

Size of Percentage Difference. Table 6.3 also reports the number of cases in each independent-variable category. There are 975 men and 1,308 women in the sample. These values are large enough to make us trust the percentages. If there were only 40 people in each category, we would have very little confidence in the percentages. The sampling error with 975 men and 1,308 women is smaller than the 6 percent difference in turnout rates. The large sample size in this table allows us to state that the 6 percent difference is real.

[1]The variable that we call "turnout" is actually based on how the respondents in the survey answered when asked about voting. There is a slight tendency on the part of respondents to report voting when they have not actually voted. Consequently, our figure for turnout in 1972 is somewhat higher than the actual turnout. However, the relationship between sex and turnout was probably not affected by this inflation, since one sex would not be likely to exaggerate its turnout more than another.

Table 6.4 Maximal Sampling Errors for Differences in Proportions

Size of Other Sample or Group	Size of One Sample or Group: 3,000	1,000	700	500	300	200
3,000	4	5	6	7	8	10
1,000		6	7	8	9	11
700			8	8	10	11
500				9	10	12
300					11	13
200						14

NOTE: The figures in the table are maximal because they represent the sampling errors for proportions in the range of 35 to 65 percent. The sampling errors decline when the proportions are more extreme, especially when the proportions are below 10 percent or above 90 percent.
SOURCE: Donald P. Warwick and Charles A. Lininger, *The Sample Survey* (New York: McGraw-Hill, 1975), p. 313. Reprinted by permission of the publisher.

How big is the relationship between sex and turnout? There could have been a 100 percent difference if all the men had voted and none of the women had voted; there could have been a 0 percent difference if men and women had voted in equal proportions. The 6 percent difference found in the table falls near the low end of this 0 to 100 continuum. It obviously shows a very small rather than a large relationship.

Actually, with survey data on attitudes, very large percentage differences are rarely found. A 60 percent difference would be enormous, and most researchers would consider a 30 percent difference large. A 5 or 10 percent difference might be the result of sampling error when the sample size is around 1,500. When sampling error is suspected, a table such as 6.4 can be consulted.[2] Even though a figure is very close to usual sampling error, we will not discount it if it fits with our previous knowledge. The fact that all the surveys up to 1972 show men voting more than women suggests that our 6 percent difference was real even though it was close to possible sampling error.

---

[2]Table 6.4 gives the maximum sampling error between two proportions based on the number of people used as the base for each proportion. If the difference is larger than the value shown in the table for the given numbers of respondents, then the difference is statistically significant. For example, with 1,000 people in each of two groups (about 1,000 men and about 1,000 women), the sampling error is no greater than 6 percent. Thus the difference shown in Table 6.3 is just barely significant. There are just 5 chances out of 100 of getting a difference this large by chance alone. The error for percentages below 35 and above 65 is smaller than shown here. More complete tables are given in Donald P. Warwick and Charles A. Lininger, *The Sample Survey* (New York: McGraw-Hill, 1975), p. 313.

Missing Data

Exactly the same problems with the missing-data categories must be faced with tables of this sort as were faced with the frequency distributions and percentages in Chapter 5. Just as before, we generally omit those categories that represent missed questions. For example, in Table 6.3, we omitted the people whose sex was not or could not be determined by the interviewer. But again the final test must be whether the missing-data category is substantively important. If it is important, it should be included. When missing data are included in a table, they are generally omitted from the percentages. However, this again rests with the substantive significance of the material.

There is a problem with missing data in bivariate distributions that did not arise with one-variable distributions. Obviously, we must have a value on each of the two variables for a respondent before that respondent can contribute to our understanding of the relationship between those two variables. If we are missing one, we must therefore exclude both. For this reason, the number of cases excluded from the table is usually nearly equal to the sum of the number of missing cases for each of the two variables.

Larger Tables

Table 6.3 is a very simple table in that both variables (sex and turnout) have only two categories. Table 6.5 shows a more complex table—the relationship between party identification and the vote in 1972. We are trying to explain the respondents' votes, so "vote" is the dependent variable. Party identification is our presumed cause

Table 6.5 Vote in 1972 by Party Identification

Vote	Strong Democrat	Weak Democrat	Independent	Weak Republican	Strong Republican
McGovern	73%	48%	34%	9%	3%
Nixon	27%	52%	66%	91%	97%
Total	100%	100%	100%	100%	100%
(Number of Cases)	(252)	(396)	(474)	(247)	(211)

SOURCE: Center for Political Studies, 1972 American National Election Study.

and is thus the independent variable. There are five categories of party identification and two categories of vote.

Because Table 6.5 is more complex than the preceding tables, it transmits more information than a two-by-two table, such as Table 6.3. Table 6.5 shows the detailed relationship between party identification and vote in 1972. Not only does it tell us that party identification affects the vote but also that there is a larger difference between strong and weak Democrats than between strong and weak Republicans. While 73 percent of the strong Democrats voted for McGovern, only 48 percent of the weak Democrats did so. The percentage vote for McGovern continues to fall as one moves to the Republican end of the spectrum: 34 percent for Independents, 9 percent for weak Republicans, and 3 percent for strong Republicans.

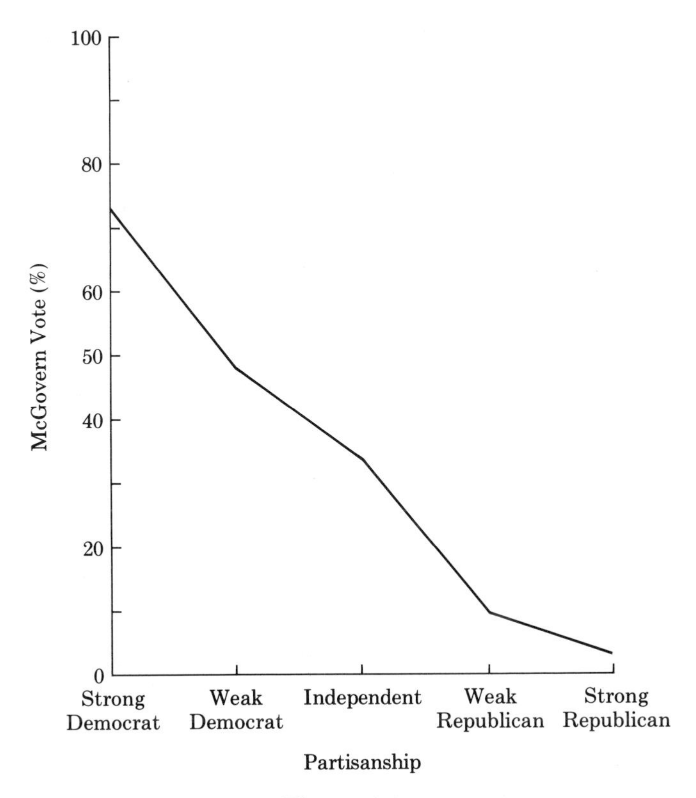

Figure 6.1
1972 McGovern Vote by Party Identification
(Based on Center for Political Studies, 1972 American National Election Study.)

Note that the difference between strong and weak Democrats is 25 points, but it is only 6 points between strong and weak Republicans. So we might conclude that not only does party affect vote but that strength of party identification in 1972 affected Democrats more than Republicans. However, these findings might be subject to revision if we assessed the impact of other variables.

Note that looking at relationships rather than simple percentages makes some difference in the impact of certain types of errors. For example, assume that in 1972 the CPS sample had twice as many Republicans as it should have had—916 instead of the 458 reported in Table 6.5. If there was no other bias in the sample, then our percentage table might not be affected *at all* by this error. The relationship might look exactly the same even if Republicans had been oversampled. However, this hypothetical oversampling of Republicans distorts the *total* percentage vote for Nixon. Instead of the 64.3 percent voting for Nixon with 458 Republicans in the sample, the figure would have been 70.9 percent with 916 Republicans. Sampling error tends to affect bivariate relationships less than it affects univariate percentage distributions.

As the tables become larger—that is, as the variables have more categories—it generally becomes increasingly difficult to summarize what is happening in the table. Graphs are sometimes employed to help in this regard, as is shown in the way Figure 6.1 presents the results of Table 6.5. Chapter 7 offers some summary measures of the strength of relationship as another way of overcoming this problem.

INTERPRETING RELATIONSHIPS

An important perspective is that relationships between variables are usually smaller than we would expect. It is best to admit this rather than to overinterpret tables.

As an example, consider survey-research evidence on attitude differences between social groups. Our stereotypes lead us to assume that everyone in a group thinks alike. In fact, attitude differences between men and women, between blacks and whites, and between young and old are usually smaller than we expect. Sex, race, and age have only limited effects on attitudes. Attitude differences as large as 10 percent are rare. Some of the largest differences on attitude

Table 6.6 Group Attitude Differences (1972 CPS National Election Study)

Question	18–24-Year-Olds	Older Than 24	Women	Men	Whites	Blacks
U.S. involvement in Vietnam was right.	36%	28%	26%	32%	30%	18%
Marijuana should be legalized rather than increasing penalties for its use.	49%	17%	19%	25%	22%	23%
Women should have an equal role with men rather than women's place being in the home.	60%	47%	47%	51%	48%	52%
Permit abortion for nonmedical reasons.	54%	40%	40%	44%	43%	34%
View protecting rights of accused as more important problem than stopping crime.	49%	33%	34%	37%	33%	56%
Favor government health insurance rather than private health plans.	49%	45%	45%	47%	42%	69%
Disapprove protest marches that are permitted by local authorities.	18%	45%	42%	39%	43%	23%
Very warm feelings toward policemen (75–100 on a scale ranging from 0–100).	31%	56%	54%	50%	55%	34%

SOURCE: Center for Political Studies, 1972 American National Election Study.

questions that we could find in the 1972 election study are shown in Table 6.6. Even most of these relationships are not all that strong; only four of them have differences of 25 percent or more. The table also shows where large differences emerge and shows that differences do not exist where we might expect them.[3] This is one of the valuable features of working with survey data—correcting our stereotypes to demonstrate that social differences have limited attitudinal effects, at least in the America of the 1970's.

To repeat, a table may show that a relationship exists, but the table should not be overinterpreted. If the differences in the table are

[3]Part of the reason the differences are not extreme is that the groups named are overlapping. For example, you might expect that men, whites, and young people would favor abortion, and that women, blacks, and older people would oppose it. But what about white women? Young blacks? Older whites? The most you can expect to find are limited tendencies among such overlapping groups.

modest, that should be admitted rather than claiming that the differences are major. Even when major differences are found, causation is not proven. The role of other variables in affecting the dependent variable must be examined before one can speak of causes. As we shall see in Chapter 8, taking additional variables into account can change the apparent importance of the original two-variable relationship. Thus, examining tables is only the beginning of statistical analysis.

Questions

Education	Position on the War			
	Dove	Intermediate	Hawk	Total
College	30%	30%	40%	100%
High school	35%	30%	35%	100%
Grade school	40%	30%	30%	100%

1. The independent variable in this table is ____________.
2. The dependent variable in the above table is ____________.
3. Are the percentages in the correct direction in the above table?
4. The 40 percent figure in the college row and hawk column means that:
   a. 40 percent of the respondents are hawks with a college education.
   b. 40 percent of the hawks have a college education.
   c. 40 percent of the college educated are hawks.
   d. 40 percent of the college-educated hawks voted.
5. Does the table suggest that those with more formal education are more dovish (antiwar) or more hawkish (prowar)?
6. Is the relationship between formal education and position on the war large or small?
7. What simple piece of information is missing from the table but is necessary to evaluate the importance of these results?

7

Measures of Association

How strong is the relationship between a pair of variables? Are the two variables associated with each other, or is there no relationship between them? The cross-tabulation shows the full relationship between ordinal or nominal variables, but it is often useful to summarize such a relationship with a single numeric value. It is particularly difficult to gauge the extent of a relationship between two variables in a table with several columns and rows; again, a summary measure is helpful. *Measures of association* are computed to summarize the degree of relationship between a pair of variables. A measure of association is a single number whose sign and size convey information about a larger cross-tabulation. (Measures of association are also called "correlation coefficients," although the latter term is sometimes used to refer to one specific measure of association, Pearson's *r*, described in Chapter 11.)

LOGIC OF MEASURES OF ASSOCIATION

Two Association Models

What is meant by "association" between variables? The usual way to think of a relationship is in terms of *covariation*—the extent to which a change in one variable is accompanied by a change in another variable. If two variables vary or change together (either in the same direction or opposite directions), we say that they covary and are associated or related. A *perfect positive relationship* is a relationship in which an increase in the independent variable is always accompanied by an increase in the dependent variable. Similarly, a *perfect negative relationship* is a relationship in which an increase in the independent variable is always accompanied by a decrease in the dependent variable. *No relationship* exists when the dependent variable is as likely to increase, decrease, or remain the same when the independent variable changes.

Another common way to think of a relationship is in terms of *predictability*—how well the category of the dependent variable can be predicted from the category of the independent variable. (This is not exactly the same as the idea of covariation, mentioned in the preceding paragraph.) In a perfect positive relationship, knowledge of the independent variable value permits perfect prediction of the value of the dependent variable. Similarly, if there is no relationship, knowledge of the independent-variable value is of no aid at all in predicting the value of the dependent variable.

In most respects the covariation model and the predictive model are the same; throughout the introductory material in this chapter, we do not draw any distinctions. Later on, we shall demonstrate the possible differences between these models (see Tables 7.7 and 7.10).

Extreme Relationships

Table 7.1 illustrates the idea of a perfect relationship for both the predictive and covariation models. The variables here are education as

Table 7.1 Perfect Relationship Between Education and Class

Class	Education		
	Grade School	High School	College
Working	100%	0%	0%
Middle	0%	100%	0%
Upper	0%	0%	100%
Total	100%	100%	100%
N (total 2,611)	(510)	(1,342)	(759)

tau-*b* = 1.00 d_{rc} = 1.00 gamma = 1.00
lambda$_{rc}$ = 1.00 tau-c = .920

NOTE: The correlation coefficients ("tau-*b*" and so forth) will be explained later in the chapter, where the reader will be referred back to this table.
SOURCE: Hypothetical.

Table 7.2 No Relationship Between Education and Class

Class	Education		
	Grade School	High School	College
Working	30%	30%	30%
Middle	50%	50%	50%
Upper	20%	20%	20%
Total	100%	100%	100%
N (total 2,611)	(510)	(1,342)	(759)

tau-b = .000 d_{rc} = .000   gamma = .000
lambda$_{rc}$ = .000   tau-c = .000

SOURCE: Hypothetical.

the predictor variable and the person's "subjective social class" (the class to which people consider themselves to belong) as the dependent variable. According to these hypothetical data: those with a grade-school education consider themselves to belong to the working class; those with high-school education consider themselves to belong to the middle class; those with a college education consider themselves to belong to the upper class. Knowing a person's educational level allows one to predict perfectly the person's class identification. As education increases, so does subjective social class. Consequently, there is a perfect relationship between education and subjective social class for both models.

By way of contrast, Table 7.2 illustrates the idea of "no relationship." Of those with grade-school education, 30 percent consider themselves to be part of the working class, as do 30 percent of those with a high-school education, and 30 percent of those with a college education. Of those with a grade-school education, 50 percent consider themselves part of the middle class, and the same holds true for the higher educational levels. Finally, 20 percent of each educational group considers itself part of the upper class. A higher education does not lead to a higher subjective social class according to these hypothetical data. Instead, education has no effect on class—people with little or no education are as likely to consider themselves members of a particular class as those with more education. Education and class are "statistically independent" here, meaning that the proportions reading down each column are identical. Knowledge of a person's education provides no added information about his or her class. Class does not vary with education. This is a case of no relationship for both models.

You would not expect education and status to be totally unrelated as they are in Table 7.2, just as you would not expect them to be perfectly related as in Table 7.1. Instead you would expect the results to be somewhere in between those shown in the two tables. But is the relationship closer to that shown in Table 7.1 or closer to that in Table 7.2?

Features of Measures of Association

The measure of association is designed to indicate how strong the relationship is—how close to perfect or null. In order to convey this information with a single number, the following conventions are usually followed:

1. The measure of association equals 1.0 (or −1.0), if there is a perfect relationship.
2. The measure equals 0.0 if there is no relationship.
3. The greater the relationship, the greater the measure's magnitude (absolute value).

One final extreme condition that was alluded to earlier is possible in the relationship between a pair of variables. Table 7.3 illustrates a hypothetical perfect negative relationship. This relationship is perfect, in the predictive sense, in that class can be predicted perfectly from education. But this relationship is negative in that *higher* education leads to *lower* subjective social class. Since a higher value on the independent variable leads to a lower value on the dependent variable, this is a perfect negative relationship in the covariation sense as well. In reality, anyone with the slightest knowledge of education and social

Table 7.3 Negative Relationship Between Education and Class

Class	Education		
	Grade School	High School	College
Working	0%	0%	100%
Middle	0%	100%	0%
Upper	100%	0%	0%
Total	100%	100%	100%
N (total 2,611)	(510)	1,342)	(759)

tau-b = −1.00 d_{rc} = −1.00 gamma = −1.00
lambda$_{rc}$ = 1.00 tau-c = −.920

SOURCE: Hypothetical.

class would expect to find a positive relationship like that in Table 7.1 (the higher the education, the higher the class), only weaker. Table 7.3 is provided merely as an example of what a negative relationship would look like.

The sign of the measure of association is used to show the direction of the relationship. The following additional conventions are followed:

4. A value greater than zero represents a positive relationship.
5. A value less than zero represents a negative relationship.

It should be obvious that the direction concept can apply only when the categories are at least ordered; thus, the distinction between positive and negative relationships is made only at the ordinal and interval levels of measurement and not at the nominal level.

Incidentally, some caution is required in dealing with the sign of the measure of association. It indicates the direction of the relationship between the variables *as coded*—and that depends on how the researchers happened to code the variables. For example, they might have set up their education variable so that the first category is college, the second is high school, and the third is grade school. The cross-tabulation might then look like Table 7.4. Is this a positive or negative relationship? The measure of association for Table 7.4 as presented will have a negative sign whether it is calculated by hand or computer. Substantively, the table still represents a positive relationship between education and class (exactly the same as Table 7.1). The measure of association is negative because the categories of the education variable have been reversed. If you verbalize the structure of the table—"greater educational background is associated with higher class identification" —you will clearly understand the table regardless of the sign of the

Table 7.4 Reversed Education Variable

Class	Education		
	College	High School	Grade School
Working	0%	0%	100%
Middle	0%	100%	0%
Upper	100%	0%	0%
Total	100%	100%	100%
N (total 2,611)	(759)	(1,342)	(510)

tau-b = −1.00 d_{rc} = −1.00 gamma = −1.00
lambda$_{rc}$ = 1.00 tau-c = −.920

SOURCE: Hypothetical.

measure. This problem is actually rather common, so it is a good idea to get in the habit of verbalizing the structure of the table rather than looking only at the measure of association. Of course, the problem can be avoided by always setting up tables so that the categories of the row variable increase as they go down the page and the categories of the column variable increase as they go from left to right. When the correlation coefficient is negative yet the substantive relationship is positive (or vice versa), either the columns or rows can be reversed and the sign changed.

Table 7.5 gives the actual relationship between education and subjective social class from the 1972 election study. In our real table there are only two class categories because only an extremely small number of people (2 out of 2,705) considered themselves "upper class," and, therefore, that category was dropped from the table.

Obviously the relationship in Table 7.5 is meaningfully positive—higher education leads to higher class identification. Yet the relationship is neither perfect nor null. How strong is the relationship? Given the five conventions we have listed, it will be some positive fraction between zero and one. According to convention 3, the number assigned for this table must be higher than the number assigned for weaker relationships and lower than the number assigned for stronger relationships, but that still does not provide an exact value. What specific value is appropriate?

Alternative Measures

The answer is that there are a wide variety of different values which different statisticians would give. Different measures of association have been developed because of different needs. The most

Table 7.5 Actual Relationship Between Education and Class in 1972

Class	Education		
	Grade School	High School	College
Working	76.9%	61.3%	29.0%
Middle	23.1%	38.7%	71.0%
Total	100.0%	100.0%	100.0%
N (total 2,611)	(510)	(1,342)	(759)

tau-b = .331 d_{rc} = .297 gamma = .567
lambda_{rc} = .271 tau-c = .364

SOURCE: Center for Political Studies, 1972 American National Election Study.

obvious reason for having different statistics available has to do with the distinctions raised in Chapter 5 among the three levels of measurement—interval, ordinal, and nominal. Different arithmetic operations are appropriate for the different levels of measurement. As a consequence, the statistical operations upon which a specific measure of association is based for one level of measurement would not be appropriate for another level. Classical statistics provides measures of association for the interval level. However, social-science data tend to be ordinal or nominal, so we will delay our treatment of interval-level measures until Chapter 11. Interval statistics were thought to be inappropriate for ordinal and nominal data, so other measures were developed for those levels of measurement, which we shall describe in the remainder of this chapter.

Also, there are a variety of measures of association because we can choose to focus on the amount of covariation, on the amount of predictive association, or on how much a change in one variable affects the other. Additionally, there are still other models of association that result in still other measures. So, we have different statistics not only for each level of measurement but also for each model or notion of relationship. (Note that several of these measures of association are provided in Table 7.5 for later reference.)

ORDINAL MEASURES

The three most commonly used measures of association at the ordinal level are "tau," "*d*," and "gamma." They have a great deal in common and usually have similar values, although each has its own purpose. Unfortunately for the person who is trying to understand these statistics for the first time, there is no simple rule that we can give to determine the appropriate statistic to use, nor can we say that one or another is always best. In fact, there is no general consensus among researchers on which measure is best. Different researchers (and instructors) emphasize different measures. Therefore, we shall explain each, although in most situations our preference is for tau.

A Covariation Measure

Kendall's tau (τ, hereafter referred to simply as "tau") is probably the most-used ordinal measure of association. It measures the extent to which an increase in one variable is accompanied by an increase in another variable (or decrease, if the sign is negative). It precisely fits the covariation model discussed earlier. Some say that it measures the correlation between the two variables. (Caution must be used, however, with respect to the word "correlation," because, as was noted earlier, it is used sometimes for a specific interval measure of association and other times for all measures of association.) Since tau measures covariation, or how things vary together, it is always calculated in exactly the same way (that is, it has the same value) regardless of which variable is the independent variable and which is the dependent. Therefore, tau is what is called a "symmetric" measure. Tau also tends to have a value that is close to that of the most important interval measure of association (Pearson's *r* correlation statistic), a fact which makes tau easier to interpret.

After we have calculated a value of tau (or a computer or another researcher has done it for us), we are faced with another question: How do we judge the importance of that value? We know that 1.0 is a perfect relationship and 0.0 is no relationship at all, but what is an acceptable value for tau? As a rough guideline, we would term tau's above .7 "high," between .3 and .7 "moderate," and between .0 and .3 "small." Ideally, we would like tau's of .9 or higher. But all of the sources of error in research tend to reduce the size of relationships, so a value of .7 is still considered high. As a result, a value of .3 can be considered reasonable—not large, but still reasonable. Actually, in some types of surveys, values as high as .3 are rare, so correlations of even .1 are reported as important. Much depends upon the state of knowledge in the field, and whether the proper predictors are known. Ideally, only correlations of .3 and higher would be emphasized, but smaller relationships can sometimes still be of interest—in such cases, it is important to recognize that the relationship is small.

There are actually two forms of the tau statistic. "Tau-*b*" can attain the value 1.0 only if there are an equal number of categories of each of the two variables (such as when there are two partisanship categories and two vote categories). "Tau-*c*" corrects for unequal numbers of variable categories and can attain the value 1.0 if there are an unequal number of independent-variable and dependent-variable categories. Note that only tau-*b* need be calculated if the variables have the same number of categories—why correct for unequal numbers of categories if they are equal? In fact, if tau-*c* is used when the variables have the same number of categories, as in Table 7.1, it will not always produce the value 1.0 for a perfect relationship. As a result, tau-*b* seems to us to be the more generally useful measure, although we do recognize that others put their faith in tau-*c* instead.

The tau-*b* value for Table 7.5 is .33, showing a moderate relationship between education and class. The tau-*c* value for this same table is .36, which is close enough to suggest a similar conclusion. Since there are an unequal number of rows and columns in the table, some would consider tau-*c* more appropriate here.

Note that we report the tau values with two digits after the decimal point. The computer often provides more digits (such as .33095), but using those digits would give a false sense of precision. Given the amount of error in any set of survey data, more than two-digit accuracy cannot be taken seriously.

A Percentage-Difference Measure

"Somer's *d*" (hereafter referred to as "*d*") is another ordinal measure of association. It is actually a direct generalization of the percentage-difference logic that we used to compare categories of the independent variable in Chapter 6. It has been developed to summarize percentage differences when variables have more than two categories. The value of *d* is an indicator of how large the change is in the dependent variable for a change in the independent variable. For the data in Table 7.5, *d* was calculated to be .30, which is a moderate relationship. (Because *d* measures the relative change in the dependent variable for changes in the independent variable, it is analogous to the regression coefficient for interval data. The regression coefficient will be discussed in Chapter 11.)

Recall that when the data in a table are presented in percentages, the percentages should be calculated within the categories of the independent variable in order for the comparisons to be related to the causal relationship under examination. Similarly, since *d* is based upon the same percentage-difference logic, it is calculated differently depending upon whether the column or the row variable is the independent variable. Statistics like *d* that have different values depending upon which variable is the independent variable are called *asymmetric measures*. Since there are two *d* values for any table, a system is needed to distinguish them: "d_{rc}" is used to indicate the value of *d* when the row variable is the dependent variable and the column variable is the independent variable; "d_{cr}" denotes the opposite—the column variable is dependent and the row variable is independent.

Computers are generally programmed to print out both d_{rc} and d_{cr}. The researcher must decide which is correct for that table. For Table 7.5, $d_{rc} = .30$ and $d_{cr} = .36$. It would be fallacious to report both values, since only one of the two variables is dependent. Here, class—the rows—is the dependent variable; hence, d_{rc} is the appropriate measure. (Incidentally, the tau-b value is always between the d_{rc} and d_{cr} values. Tau-b for Table 7.5 is .33, clearly between $d_{rc} = .30$ and $d_{cr} = .36$.)

A Measure for Scale Relationships

Goodman and Kruskal's gamma (γ) statistic is yet another measure of association for ordinal variables; it measures scale-type relationships. Instead of measuring independent-variable effects on dependent variables, we might want to know whether two variables "scale," that is, whether they both measure the same underlying dimension. Scaling will be considered more directly in Chapter 9, but we can at least illustrate the topic here. Suppose we suspect that interest in the political campaign and concern over the election outcome are really two measures of the same underlying dimension, which we might call "political involvement." We hypothesize that: People who are not at all involved in politics would show low interest and low concern; people with great involvement would show high interest and high concern; people with medium involvement might show high concern but low interest. It is hard to imagine how anyone could be highly interested in the campaign if they had low concern about the outcome (except perhaps a foreign visitor). Table 7.6 illustrates this hypothesis. If this hypothesis is correct, concern and interest are both measures of involvement, but interest is a stronger indicator than concern—people with medium involvement can show high concern, but only those people with high involvement can show high interest. Notice that this leaves one cell of the table empty. Gamma seeks to measure this type of relationship. Gamma is 1.0 for Table 7.6 because there is a perfect scale relationship between interest and concern (with only high-concern types showing high interest).

One of the most important difficulties with gamma is that it tends to produce high values, often much higher than tau. In fact, gamma is

Table 7.6 Scale Relationship Between Interest and Concern

Concern Over Outcome	Interest in Election		
	Low	High	
Low	30%	0%	
High	40%	30%	
Total			100% (1,000)

tau-b = .429 d_{rc} = .429 d_{cr} = .429 gamma = 1.00

NOTE: This table has been percentagized "on the corner" or "by the total" so that the sum of all the percentages in the table is 100. This is particularly appropriate where there is no distinction between independent and dependent variables.

SOURCE: Hypothetical.

never smaller than tau. In Table 7.6, note that tau-*b* is only .43 and that gamma is 1.0. This feature of gamma makes it attractive to researchers who want to report strong relationships. They will have more to report if they use gamma instead of tau. This is meant not as a suggestion but as a warning: Beware of gammas, since they may be overstating a relationship. Some researchers argue that gamma is a measure of predictive association for ordinal data and would, therefore, prefer gamma to tau. Yet, because of the prediction rule used by gamma[1] and because of the special circumstances under which gamma can have a value of 1.00, we do not recommend its use except for the scale-type situations described above.

Choice of Measure

Now, which of the measures of association should be used for ordinal variables? Tau, *d,* or gamma? One would generally be correct in using tau. One might use *d* instead, but tau is used more generally. Use

[1]In comparing two ordinal variables, gamma measures how often knowing the positions of respondents with regards to one variable would allow you to predict the positions of respondents on the second variable. Numerically, it is the number of correct predictions (that is, the numbers of pairs of respondents with the same order on the two variables) minus the number of incorrect predictions, that difference divided by the total number of predictions. In this sense, gamma indicates the amount that the error was reduced by knowing the people's responses on the independent variable. For this argument, see Herbert L. Costner, "Criteria for Measures of Association," *American Sociological Review, 30* (1965), 341–53. The view in the text is developed in Herbert F. Weisberg, Models of Statistical Relationship," *American Political Science Review, 68* (1974), 1638–55.

gamma only if your model of a perfect relationship is the scale-type relationship. The most important thing is to understand the table; know what it is saying. Once you understand the table, you will probably want to use a measure of association to summarize the amount of relationship in the table. Remember that understanding the table is most important, and statistics should aid in that understanding rather than substitute for it.

Before we leave these ordinal measures, we should note that although they all do not agree on what constitutes a perfect relationship, they do agree on what constitutes no relationship (that is, if one ordinal measure were zero, the others would also be zero). Also, they would all have the same sign for any particular table.

Use of Ordinal Statistics

Not all lists of the categories of a variable are in a meaningful order. Some categories may be out of order; some categories cannot be ordered—and are thus not ordinal. For example, say that vote were divided into the categories of "voted Democratic," "voted Republican," and "did not vote." The "did not vote" category prevents this from being an ordered set of categories. After all, voting does not range from voting Democratic at one extreme to not voting at the other, with voting Republican somewhere in the middle. Ordinal statistics using this set of categories would not be meaningful. If vote were categorized in that way, the "did not vote" category would have to be removed, and the statistics would have to be computed using only Democratic and Republican voters. Only then would ordinal statistics be meaningful. More generally, "missing data" must always be excluded before computing any measures of association. Also remember that variables with only two categories may be considered either nominal or ordinal. Sex may not have two orderable categories, but that does not make any difference; because sex has only two categories, it can be considered ordinal or nominal for purposes of choosing measures of association for tables.

So far we have stressed using ordinal measures for ordinal variables. All ordinal measures of association assume that as one variable increases the other will consistently increase or consistently decrease. Difficulties arise when the data do not conform to such a monotonic relationship, that is, if the second variable increases for a while and then

Table 7.7 Predictive Relationship Between Education and Class

Class	Grade School	High School	College
Working	100%	0%	0%
Middle	0%	0%	100%
Upper	0%	100%	0%
Total	100%	100%	100%
N (total 2,611)	(510)	(1,342)	(759)

tau-b = .025 d_{rc} = .025 gamma = .025
lambda$_{rc}$ = 1.00 tau-c = .023

SOURCE: Hypothetical.

decreases, or vice versa. In Table 7.7, we show a relationship in which the independent variable, education, is not monotonically related to the dependent variable. The covariance measure, tau-b, for this table is nearly zero, as is gamma. However, we can perfectly predict the value of the dependent variable from the value of the independent variable, as the 1.00 value of the nominal-level measure lambda shows. (Lambda will be explained in detail in the following section.)

The important point to note in this example is that one should not blindly apply ordinal-level measures to ordinal variables if the expected relationship is not monotonic. For example, if we were to examine the relationship between age and turnout, we would not expect to find a monotonic relationship. Our expectation would be that turnout would increase with age until about retirement age, when it would begin to decrease with age. In this case we would not want to apply ordinal-level (or interval-level) measures, not because the categories of the variables were not ordered, but because the expected relationship was not monotonic.

A Measure of Correlation for Rank Orders

One additional ordinal measure is useful for a particular type of data. Some ordinal variables are "rank orders," such as the rank order of people from richest to poorest or from most conservative to most liberal. "Spearman's rho" (r_s) is used to measure the correlation between such rank orders. It is essentially the Pearson's r correlation (see Chapter 11) between the ranks. Say that, in Table 7.8: 1 stands for highest income and 5 for the lowest; 1 stands for most conservative

Table 7.8 Calculation of Spearman's Rho

Person	Income	Ideology	Difference	Difference-squared
A	1	1	0	0
B	3	2	1	1
C	2	4	−2	4
D	4	3	1	1
E	5	5	0	0
Sum			0	6

N = number of people = 5

$$r_s = 1 - \frac{6(\text{sum of squared differences})}{N(N^2 - 1)} = 1 - \frac{6(6)}{5(5^2 - 1)}$$

$$= 1 - \frac{36}{120} = .70$$

SOURCE: Hypothetical.

and 5 for the least conservative. Spearman's rho is calculated as shown in Table 7.8. The data show a strong tendency for people with more income to be more conservative, though many more than five people would be required to verify this relationship. Spearman's rho is often useful, but it requires complete rankings of people on the variables. When, instead, the data consist of ordered categories of people (such as categories of conservatives, moderates, and liberals), then Kendall's tau as presented above is more useful.

NOMINAL MEASURES

If there are more than two categories and they cannot be ordered, then there is no choice but to consider the variable nominal and apply one of the nominal-level measures of association.

A Predictive Measure

Probably the most useful nominal measure of association is lambda (λ). Lambda is asymmetric (it has different values depending on whether the row or column variable is the independent variable), and it conforms to the predictive model of a relationship. It also has a very direct empirical interpretation. The value of lambda is the proportion that error in predicting the value of the dependent variable is reduced by knowing the value of the independent variable (that is, how much better is your prediction of the dependent variable if you know the value of the independent variable). Measures of association that have this property are called PRE (*p*roportional *r*eduction in *e*rror) measures.

Table 7.9 is an example of the use of lambda. Note that we have included raw frequencies as well as the percentages so that we can demonstrate some calculations. If we did not know a person's race (independent variable), how would we guess that person voted in 1968? Well, more people voted against Humphrey (509) than voted for him (418), so we would be correct more often if we guessed "against Humphrey." How often would we be wrong? We would be wrong 418 times out of 927. The proportion of errors would be 418 ÷ 927. If we knew the person was black, how would we guess he or she voted? "For Humphrey," of course. How often would we be wrong? We would be wrong 5 times out of 89. If we knew a person was white, we would guess he or she voted against Humphrey, and we would be wrong 334 times out of 838. Hence, if we knew the value of the independent variable we would have been wrong 339 (5 + 334) times out of 927. The proportion of error if we knew the independent variable was 339 ÷ 927. Lambda, the proportion that we reduced our original error, is equal to the original proportion of error (418 ÷ 927) minus the proportion of error if we know the value of the independent variable (339 ÷ 927), all divided by the original proportion of error (418 ÷ 927),

$$\lambda = \frac{\dfrac{418}{927} - \dfrac{5 + 334}{927}}{\dfrac{418}{927}} = \frac{418 - 339}{418} = \frac{79}{418} = .189$$

Another way to think of lambda is that it measures the improvement in prediction when the independent variable is known. The number of errors we would make not knowing a respondent's race is 418 errors. However, we would make only 339 errors (5 + 334) if we knew his race; hence, we would reduce the number of errors by 79. That number would be all right except that, for the sake of making comparisons, we like our measures of association to go from zero to one. If we divide the improvement (79) by the maximum that we could have improved (we could

Table 7.9 Presidential Vote by Race in 1968

Vote	Black		White		Total	
	Percent	Frequency	Percent	Frequency	Percent	Frequency
For Humphrey	94%	84	40%	334	45%	418
Against Humphrey	6%	5	60%	504	55%	509
Total	100%	89	100%	838	100%	927

$N = 927$

tau-b = .323 d_{rc} = .545 gamma = .924 lambda$_{rc}$ = .189

SOURCE: Center for Political Studies, 1968 American National Election Study.

have eliminated all 418 errors), we then have a measure that varies between zero and one.

Dichotomous Variables

Since race and vote are dichotomous variables, it is possible to calculate the several ordinal statistics for them. They are presented in Table 7.9. The value of d_{rc} should be easy to interpret and understand here (94 percent − 40 percent = 54 percent difference). Gamma is very high because there were so few blacks who voted against Humphrey; if no blacks had voted against Humphrey, gamma would have been 1.0. However, gamma is certainly not appropriate to the type of relationship we are examining. One could make an argument for using tau-b or d_{rc} here, but the interpretation would be significantly different than it is for lambda. Remember that the signs of correlation coefficients for nominal variables are not substantively meaningful; for example, in Table 7.9 it would not make sense to call the relationship between race and vote either positive or negative.

Choice of Measures

One of the largest contrasts between predictive measures like lambda and the covariation measures like tau is illustrated in Table 7.10. Lambda is zero here because knowing the value of the independent variable does not improve our ability to predict the value of the dependent variable. Regardless of the category of the independent variable, we always guess that the person voted for Nixon. However, the covariation measure, tau, shows that the two variables vary together to a certain degree (.39). This illustrates that you have to be clear in your own mind as to what type of model relationship you are looking for. The statistics are often quite similar; however, as Table 7.10 shows, they can also be quite different.

INTERPRETING ASSOCIATION

Significance Tests

How do you judge the importance of a relationship? One possibility would be to consider the sampling notions when gauging its importance. If we have interviewed only a sample of a population and if we want to make inferences about the population from the sample, we must take into consideration the possibility that our sample results will not hold for the population (that is, that they occurred by chance because we took a sample rather than interviewing the entire population). In dealing with a single variable, we could handle the problem by considering the sampling error of the variable. In dealing with more than one variable, we could seek to put a similar interval around the correlation value, which would tell us the range of probable values of the true relationship in the population. The size of the interval would depend upon the sample size and also upon the magnitude of the correlation. What we would be most concerned about, of course, is accidentally claiming there is a relationship when there is no relationship

Table 7.10 Vote by Party in 1972

Vote	Republican		Non-Republican		Total
	Percent	Frequency	Percent	Frequency	
Nixon	94%	429	52%	585	1,014
McGovern	6%	29	48%	537	566
Total	100%	458	100%	1,122	1,580

$N = 1,580$

tau-b = .393 d_{rc} = .415 gamma = .863 lambda$_{rc}$ = .00

NOTE: Independents and Democrats are included in the "Non-Republican" column.
SOURCE: Center for Political Studies, 1972 American National Election Study.

in the population. The question then is whether the relationship is "statistically significant."

The Chi-Square Test. There are significance tests for most of the measures described in this chapter, but we shall use as an example the best-known statistical significance procedure: the chi-square (χ^2) test. It hypothesizes that there is no relationship between the variables in the population (for an example of statistical independence, see Table 7.2) and tests whether any apparent relationship in the sample is due to chance. Consider again the relationship between education and class. In 1972, 55 percent of the sample considered themselves working class compared to 45 percent middle class. If education did not affect class status, 55 percent of each education group would consider themselves working class. There were 510 people with grade-school education, and 280 would then be expected to be working class, which compares to an observed 392 (see Table 7.5). Chi square contrasts the observed and expected values for each combination of education and class. It squares the difference between expected and observed values, divides by the expected value (for example, $[392 - 280.29]^2/280.29 = 44.52$ for grade-school working class), and sums up the quotients for all the cells of the table. Such a calculation for Table 7.5 gives a chi-square value of 328, as shown in Table 7.11.

Table 7.11 Calculation of Chi Square for Table 7.5

Category	Observed	Expected		$\frac{(\text{Observed} - \text{Expected})^2}{\text{Expected}}$
Grade school, working class	392	55% of 510 =	280.29	44.52
Grade school, middle class	118	45% of 510 =	229.71	54.32
High school, working class	823	55% of 1342 =	737.56	9.90
High school, middle class	519	45% of 1342 =	604.44	12.08
College, working class	220	55% of 759 =	417.14	93.17
College, Middle class	539	45% of 759 =	314.86	113.69
Total	2,611		2,611	$327.68 = \chi^2$

Degrees of freedom = (Number of rows − 1) × (Number of columns − 1)
= (2 − 1) × (3 − 1) = 1 × 2 = 2

$$\chi^2 = \sum \frac{(\text{Observed} - \text{Expected})^2}{\text{Expected}}$$

Table 7.12 Chi-Square Values Required for Significance (Permitting a 5% Chance of Error)

Degrees of freedom	Chi Square	Degrees of freedom	Chi Square	Degrees of freedom	Chi Square	Degrees of freedom	Chi Square	Degrees of freedom	Chi Square	Degrees of freedom	Chi Square
1	3.84	6	12.59	11	19.68	16	26.30	30	43.77	80	101.88
2	5.99	7	14.07	12	21.03	17	27.59	40	55.76	90	113.14
3	7.81	8	15.51	13	22.36	18	28.87	50	67.50	100	124.34
4	9.49	9	16.92	14	23.68	19	30.14	60	79.08		
5	11.07	10	18.31	15	25.00	20	31.41	70	90.53		

SOURCE: *Biometrika Tables for Statisticians,* vol. 1, 3rd ed. Edited by E. S. Pearson and H. O. Hartley, Table 8. (New York: Cambridge University Press, 1966.) Reprinted by permission of the Biometrika Trustees.

Is it likely that such a value (or larger) would have occurred by chance? The "degrees of freedom" for a table is defined as: (one less than the number of rows of the table) × (one less than the number of columns), or $(2 - 1) \times (3 - 1) = 1 \times 2 = 2$, for our example. Table 7.12 shows the chi-square value required for significance for different numbers of degrees of freedom, taking no more than a 5 percent chance of concluding that a relationship is significant when in fact it is not. With 2 degrees of freedom, a chi-square value of 6 or more will occur by chance less than 5 times out of 100. Since we found a chi square of 328, we can confidently reject the hypothesis (often termed the "null hypothesis") that there is no relationship between the two variables. The relationship is greater than would be expected on the basis of chance and sampling error. Note that chi square is essentially a test for a predictive relationship at the nominal level; it does not check whether respondents who are higher on one variable are higher on the other.

Problems with Significance Tests. Unfortunately, researchers often face two problems in dealing with statistical significance. First, most tests of statistical significance (such as the popular chi square) are based upon the assumption that the sample is a simple random sample. When other types of samples are used, the assumptions of these significance tests are not met. Second, even if the assumptions of these tests or other modified tests are met, weak relationships are found to be statistically significant when samples are large, such as those for national surveys with a typical 1,500 respondents. Thus, except for very small samples (especially those of size 100 or less), tests of statistical significance are not very discriminating.

Additional Criteria

The more important question generally is "substantive significance"—whether the relationship is large enough to be significant substantively. A correlation of .09 might be statistically significant, but it just is not very important, since it indicates that the independent variable is not of much use in predicting the dependent variable's value. We feel that substantive significance is more important than statistical significance for large samples, so that an emphasis on the significance test would give a false sense of security.[2]

[2]For additional readings on this subject, see Denton E. Morrison and Ramon E. Henkel (Eds.), *The Significance Test Controversy* (Chicago: Aldine-Atherton, 1970).

In any case, we should not just look for the largest correlations. That strategy would take advantage of all the chance variation in the data, and just by chance there will be a few large correlations. Rather than seeking the largest correlations, we should be guided by *theory* in our choice of which correlations to examine. If those correlations turn out to be small, that is a substantively important negative finding.

Finally, our discussion of causation in Chapter 3 must be recalled Finding a relationship that is substantively important does not mean that we have "proven" that one variable "causes" the other. Correlation does not prove causation. That two variables covary does not in itself show that a change in one would produce a change in the other. There is always the possibility that some other variable or variables are causing both of the original variables to change. In order to speak more definitely about causation, we must examine more than two variables at a time. This is the subject of the next chapter.

Questions

How would you interpret the following situations? (Watch for fallacies in the applications of the statistics.)

1. The correlation between race and interest in politics is −.30.
2. The tau correlation between religion and attitudes on abortions is .15.
3. The lambda correlation between region and vote is .20.
4. The gamma correlation between attitudes toward abortion and divorce is .85.
5. The tau correlation between age and attitudes toward divorce is .05.

8

Statistical Controls

If we are interested in explaining political attitudes or behavior, as described in Chapter 3, simple two-variable tables like those that we have been looking at in the past few chapters usually do not tell us enough. For example, in Chapter 6 we examined the relationship between sex and turnout. Table 6.3 showed a weak relationship, with males turning out to vote slightly more than females. However, we were unable to offer any reasons for that difference based upon that single table. Few would contend that there is a genetic difference between men and women that causes women to vote less than men. So, the discovery of a sex difference should whet our appetites to discover its causes. Social scientists are seldom satisfied with the answers provided by simple two-variable tables no matter what the topic. Inevitably, one or more additional variables must enter into the analyses in order to explain attitudinal differences or to help determine whether the original relationship holds under tests of alternative explanations.

The effects of introducing a third variable into the analysis can be unusual. Obviously, including another independent variable may help explain the dependent variable better. What may be less obvious is that, because of the way the new variable is related to the original pair of variables, taking account of the third variable may fundamentally alter our understanding of the original two-variable relationship. An effect which seemed to be due to one variable may disappear, while the new variable may prove to be the basic causal variable. A common

Table 8.1 Fire Damage by Number of Fire Trucks at the Scene

Damage	Number of Trucks			
	None	1–2	3–4	5+
$10,000 or less	98%	40%	12%	2%
$10,001 to $100,000	2%	39%	26%	14%
$100,001 to $1,000,000	0%	21%	48%	65%
More than $1,000,000	0%	1%	14%	19%
Total	100%	100%	100%	100%
N (total 943)	(125)	(217)	(341)	(260)

tau-*b* = .570 d_{rc} = .563 gamma = .750 lambda_{rc} = .288

Source: Hypothetical.

technique for such analysis of nominal and ordinal variables is called "statistical controlling." It is a method for holding constant a third variable while examining the relationship between two other variables. This chapter deals with that technique.

HOW TO IDENTIFY SPURIOUS RELATIONSHIPS

Probably the best way to begin is to look at a relationship that nearly everyone would suspect of being inadequate to explain what is happening in the real world. Table 8.1 shows the relationship between the number of fire trucks sent to a fire and the amount of damage done by the fire (measured in dollars). Note that there is a strong positive relationship between these two variables (tau-*b* = .57). Naturally, we are not convinced that the fire trucks caused the damage.

In this example there clearly has to be some reason that makes it appear as though the fire trucks caused the damage. If we think about it for a moment, it occurs to us that the severity of the fire is related both to the amount of damage and to the number of fire trucks at the scene. Any fire chief is going to send more trucks to the more serious fires, and of course, the more serious the fire, the more dollar damage is likely to be done.

If we somehow were able to take into account the impact of the severity of the fire, we could get a better idea of the relationship between the number of fire trucks and the damage. Our expectation is, of course, that there would no longer be any relationship at all. The

Table 8.2 Fire Damage by Number of Fire Trucks at the Scene for Minor Fires Only

Damage	Number of Trucks			
	None	1–2	3–4	5+
$10,000 or less	98%	97%	97%	0%
$10,001 to $100,000	2%	3%	3%	0%
$100,001 to $1,000,000	0%	0%	0%	0%
More than $1,000,000	0%	0%	0%	0%
Total	100%	100%	100%	0%
N (total 230)	(125)	(76)	(29)	(0)

tau-b = .017 d_{rc} = .005 gamma = .097 lambda_{rc} = .000

SOURCE: Hypothetical.

question is how to accomplish this. Since empirical experiments of the type used by physical scientists are out of the question, we must devise ways of controlling for additional variables using what are called statistical controls.

What we should do is look at the relationship between the number of fire trucks and the damage for minor fires, then look at the relationship between those same variables for moderately serious fires, and finally look at the relationship for major fires. We would expect no relationship between the number of fire trucks and the damage for any of the three tables. Tables 8.2 through 8.4 show what this might look like. Keep in mind that the data in Tables 8.2 through 8.4 are exactly the same as were in the original, Table 8.1. The only thing that has been done is to divide the cases into three groups depending upon a

Table 8.3 Fire Damage by Number of Fire Trucks at the Scene for Moderate Sized Fires Only

Damage	Number of Trucks			
	None	1–2	3–4	5+
$10,000 or less	0%	10%	10%	9%
$10,001 to $100,000	0%	70%	66%	67%
$100,001 to $1,000,000	0%	20%	24%	22%
More than $1,000,000	0%	0%	0%	2%
Total	0%	100%	100%	100%
N (total 304)	(0)	(117)	(133)	(54)

tau-b = .042 d_{rc} = .036 gamma = .075 lambda_{rc} = .000

SOURCE: Hypothetical.

Table 8.4 Fire Damage by Number of Fire Trucks at the Scene for Serious Fires Only

Damage	None	1–2	3–4	5+
$10,000 or less	0%	0%	0%	0%
$10,001 to $100,000	0%	0%	0%	0%
$100,001 to $1,000,000	0%	75%	74%	78%
More than $1,000,000	0%	25%	26%	22%
Total	0%	100%	100%	100%
N (total 409)	(0)	(24)	(179)	(206)

tau-*b* = −.046 d_{rc} = −.037 gamma = −.102 lambda_{rc} = .000

SOURCE: Hypothetical.

third variable (the seriousness of the fire, measured by the number of alarms). If you add up the cases in Tables 8.2, 8.3, and 8.4, you will once again have Table 8.1. Controlling is not magic but only a different way of looking at the same thing.

For minor fires (Table 8.2), the relationship between the number of fire trucks sent and the damage has effectively disappeared. Regardless of the number of trucks sent, the damage was nearly always in the $10,000-or-less category. Also, all of the correlation measures are zero, indicating no relationship is present in the table.

Table 8.3 shows that for moderate fires, the relationship between the number of trucks sent and the amount of damage has disappeared. The distribution of the cases in damage categories is almost exactly the same for all of the categories of the independent variable (that is, the independent variable is not affecting the dependent variable); hence, no relationship is present. The correlation between the two variables is effectively zero.

Table 8.4, for serious fires, is very similar to the previous table except that the cases have moved down to the higher-damage categories. But again there is no impact of the independent variable on the dependent variable, because the distributions are essentially the same for the categories of the independent variable. Also, the correlation coefficients are again all approximately zero, indicating no relationship.

Not so surprisingly, we see that there is no relationship between the number of trucks and the damage for any of these three tables. Note that all of the tau's are near zero. Thus we conclude that the original relationship between number of trucks and damage (without controls) was *spurious*. It can be accounted for entirely by controlling for the seriousness of the fire.

HOW TO INTERPRET CONTROLS

Possible Results

How do we interpret the various patterns of correlations in control tables? If the correlations were all zero after controlling, as in the above example, then the original relationship is termed spurious. The original independent variable did not cause the dependent variable; instead, the control variable affected them both and induced an apparent relationship between them.

If the correlations in the separate control tables were considerably reduced but still above approximately .10, we could say that some of the original relationship was spurious, but not all of the variation could be explained by the control variable. On the other hand, if the original relationship remained unchanged by controlling (if the correlation coefficients were just about as high in the control tables as in the original), we merely would say that the control had no impact on the original relationship.

Still another possibility is that the relationship between the original pair of variables is different for the different categories of the control variable. The relationship might (1) be stronger for some category or categories of the control variable, (2) disappear—be zero—for some, and possibly (3) be in the opposite direction (the correlation changes its sign) for others. In these cases, we say that the original relationship is *specified* by the control variable. In other words, the control variable determines what the relationship will be like.

An Example

A common control variable in survey analysis is education; often, two variables seem to be related to one another, but both prove simply to be related to a third: education. The fact that education is related to both variables creates an apparent but spurious relationship between the first two variables. Returning to our sex and turnout

Table 8.5 Turnout Rates by Sex

	Sex	
	Male	Female
Voted	80%	69%
Did not vote	20%	31%
Total	100%	100%
N (total 1,954)	(883)	(1,071)

tau-b = .12 d_{rc} = .11
gamma = .28 lambda$_{rc}$ = .00

SOURCE: Hypothetical.

example, it is reasonable to expect that people with more education vote more than people with less education. There is a correlation between education and turnout. And in our sexist society, it is reasonable to expect that men are more likely to have had a college education than women; that is, there is a correlation between sex and education. Thus, since high education is associated both with high turnout and being male, the apparent relationship between sex and turnout *might* just be due to their both being related to education.

The concept of controlling for a third variable is important enough to restate the logic in another fashion. We found turnout differences between men and women. However, we suspect there are educational differences between men and women. Furthermore, it might well be that the turnout differences merely reflect educational differences. If we looked only at men and women who possessed the same amount

Table 8.6 Turnout Rates by Sex for Different Educational Levels

	Grade School		High School		College	
	Male	Female	Male	Female	Male	Female
Voted	60%	60%	75%	75%	90%	90%
Not Vote	40%	40%	25%	25%	10%	10%
Total	100%	100%	100%	100%	100%	100%
N (total 1,954)	(150)	(571)	(300)	(350)	(433)	(150)
	tau-b = .00		tau-b = .00		tau-b = .00	
	gamma = .00		gamma = .00		gamma = .00	
	d_{rc} = .00		d_{rc} = .00		d_{rc} = .00	
	lambda$_{rc}$ = .00		lambda$_{rc}$ = .00		lambda$_{rc}$ = .00	

SOURCE: Hypothetical.

Table 8.7 Reported Turnout by Sex for 1972

	Male	Female
Voted	76%	70%
Did not vote	24%	30%
Total	100%	100%
N (total 2,283)	(975)	(1,308)

tau-*b* = .07 d_{rc} = .06 gamma = .16 lambda_{rc} = .00

SOURCE: Center for Political Studies, 1972 American National Election Study.

of education (so there are no educational differences between them), we might find no turnout differences between them. First we would look at men and women with high levels of education, then at men and women with medium education, and then at those with low education.

Table 8.5 gives a hypothetical relationship between sex and turnout uncontrolled. There is an 11 percent difference in turnout rates, with men voting 11 percent more than women. The value of tau-*b* is .12. Although that is not very large, it does indicate that there is some relationship between sex and turnout.

Table 8.6 shows exactly the same hypothetical data broken down into three groups based upon education. Controlling for education completely eliminates the relationship between sex and turnout. Turnout rates of men and women are seen to be identical for the college educated, the high-school educated, and the grade-school educated. For each education category, the percent difference in turnout between men and women is 0 percent. The correlation between sex and turnout within each education category is therefore .00. The original relationship was spurious: There is no relationship between sex and turnout once education is controlled. Yet these control tables

Table 8.8 Reported Turnout by Sex for 1972, Grade-School Education Only

	Male	Female
Voted	67%	50%
Did not vote	33%	50%
Total	100%	100%
N (total 440)	(199)	(241)

tau-*b* = .17 d_{rc} = .17 gamma = .34 lambda_{rc} = .00

SOURCE: 1972 Center for Political Studies, 1972 American National Election Study.

Table 8.9 Reported Turnout by Sex for 1972, High-School Education Only

	Male	Female
Voted	74%	68%
Did not vote	26%	32%
Total	100%	100%
N (total 1,145)	(413)	(732)

tau-*b* = .06 d_{rc} = .05 gamma = .13 lambda_{rc} = .00

SOURCE: Center for Political Studies, 1972 American National Election Study.

are consistent with the original table—how could there be an overall turnout difference of 11 percent but zero differences within each education category? Those with low education tend to turn out less and women are overrepresented in the low-education category. Consequently, it appears as though women vote less than men when one looks at the composite table (Table 8.5). The apparent turnout differences between men and women really prove to be educational differences between the sexes.

Let us turn now from hypothetical data to real data. Real data are seldom as clear or as straightforward as hypothetical data. Table 8.7 shows the uncontrolled relationship between sex and turnout for 1972. The three following tables show the relationship controlled for education. (All of the tables are based upon the actual data from the CPS 1972 American National Election Study.)

The turnout difference between men and women with low education is 17 percent (Table 8.8); the difference was only 6 percent for Table 8.7.

The turnout difference between men and women with high-school educations (Table 8.9) more closely resembles the weak relationship in the original table; there is about a 6 percent difference in both Tables 8.7 and 8.9.

Table 8.10 Reported Turnout by Sex for 1972, College Education Only

	Male	Female
Voted	85%	89%
Did not vote	15%	11%
Total	100%	100%
N (total 696)	(362)	(334)

tau-*b* = −.06 d_{rc} = −.04 gamma = −.16 lambda_{rc} = .00

SOURCE: Center for Political Studies, 1972 American National Election Study.

The final control table for those in the high-education category (Table 8.10) clearly indicates that education specified the relationship between sex and turnout. The relationship is moderate for those with low education, weak for those with medium education, and negative (the correlation coefficient changes its sign) for those with high education. In 1972, women were less likely to vote than men, but this is true only for those people without college education and mainly for those with only grade-school education. College education proved to be the equalizer—college-educated women tended to vote more than college-educated men. College education apparently compensates for the sexual biases in political socialization in the United States. Although we still may not know all there is to know about this relationship, we certainly know more than when we had only the bivariate table.

OTHER WAYS TO LOOK AT THREE-VARIABLE RELATIONSHIPS

Depending on exactly what we are studying, we might examine the three-variable relationship in ways that differ from what we did above.

Controlling in the Opposite Direction

While controlling for education, we have examined the effect of sex on turnout, but we might instead have been interested in the relationship between education and turnout, controlling for sex. Comparing men in Tables 8.8 through 8.10, we would find that men with more education voted more often in 1972 than those with less education. A similar effect holds for women; in fact, education had a greater effect on turnout for women than for men. Again the relationship has been specified, but the point is that we might be interested in controlling either on education or on sex. Which control is employed depends

Table 8.11 Reported Turnout by Sex and Education for 1972

Education	Male		Female	
	Percent	Frequency	Percent	Frequency
Grade School	67%	199	50%	241
High School	74%	413	68%	732
College	85%	362	89%	334

SOURCE: Center for Political Studies, 1972 American National Election Study.

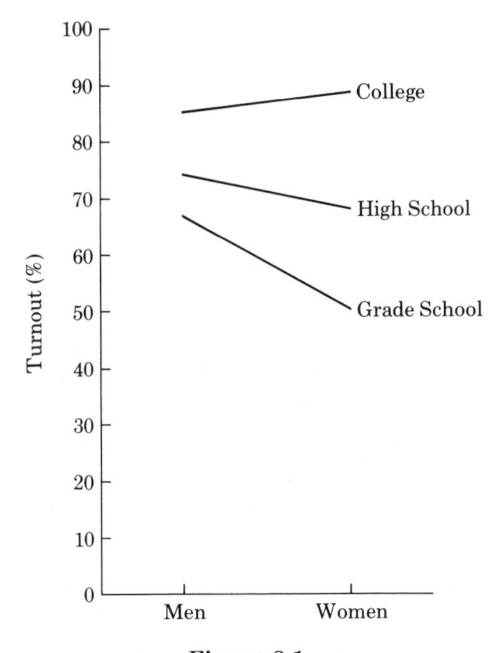

Figure 8.1
1972 Reported Turnout
By Sex and Education
(Based on Center for Political Studies, 1972 American National Election Study.)

upon our theories and on what is being studied: (*a*) sex differences and the conditions that govern them or (*b*) education differences and the conditions that govern them.

Measuring Effects of Two Variables

Another goal of the analyst might be to determine which of the explanatory variables has a greater effect on the dependent variable. The control data in Tables 8.8–8.10 need only be reorganized to highlight whether sex or education has a greater effect on turnout. Table 8.11 and Figure 8.1 show the turnout rates for different combinations of sex and education. (The number beneath each percentage in Table 8.1 is the number of cases on which it is based according to Tables 8.8–8.10.) Clearly, education differences had a greater effect on turnout in 1972 than did sex differences. However, different combinations of sex and education groups have different effects, with sex having different effects on turnout for different educational levels.

This is often termed a case of "interaction" in which the variables have not only separate effects on turnout but combined effects.

USE OF ADDITIONAL CONTROLS

Social-science data analysis generally involves using more than a single control variable, either trying a variety of separate controls or employing several controls together. If a control for education did not affect the original relationship between sex and turnout, then we might have to think more about why men might turn out more than women. For example, we might think that age is involved: young mothers might turn out less than young fathers, but otherwise there would be no turnout differences between men and women. To test this, we might check whether men and women over thirty turn out in equal proportions. We would be on the right track if they do (and we would then make further tests), but if not, we might look even harder for another possible third variable explaining the overall turnout differences. Eventually, we might decide that some variable not in the data plays a critical role in causing women to vote less often than men. We might hypothesize that women vote less because as young girls they are socialized to consider politics more men's work. We cannot test that hypothesis with the available data, but at least we would have eliminated plausible alternative explanations, and we have developed a reasonable nypothesis.

However, keep in mind that at some point one has to stop controlling even if there are thousands of possibilities. At the same time, there has to be an awareness that the original relationship may be spurious due to some variable that the analyst has not even considered. That is why we never prove theories, but only fail to disprove them.

Another consideration is that several variables might be involved in an explanation of why women vote less than men. Although we have only shown examples of one control variable, it is possible to control for more than one variable at a time. For example, we could control the relationship between sex and turnout for both (a) interest in politics and (b) education. If we did so, we would have as many tables as there are pairs of values on the two control variables. Although the logic of this analysis could be extended to any number of simultaneous control variables, the number of tables that must be constructed increases geometrically. Hence, there is some practical limit to this approach.

It is always difficult to decide which variables to control. There are several things that can be kept in mind that will help in that decision. First, control for only those variables that you know are related to both your independent and dependent variables. Controlling for a variable cannot reduce the original correlation unless it is correlated with both. Second, look to your theory and the purpose of your research for guidance. What variables would the theory suggest might be important? In particular, what variables might precede (and therefore be able to cause) the independent variable?

We have seen in this chapter why looking at two-variable relationships is only the beginning of data analysis. The apparent effect of one variable on another can change considerably when other variables are added to specify the causal process more completely. We shall return to the topic of statistical controls in Chapter 11, where we shall approach this problem from a slightly different perspective and with interval measures.

Questions

1. Tables 8.8 through 8.10 show the effect of sex on turnout rates, controlling for education. Use those tables (or Table 8.11) to construct control tables showing the effects of education on turnout, controlling for sex.
2. The 1972 Center for Political Studies election survey measured attitudes on the proper role for women. Table 8.12 shows how age affects whether

Table 8.12 Attitudes Toward Women's Role by Age

Role for Women	Age		
	Under 35	35–55	Over 55
Equal Role	54.8%	45.9%	44.7%
Neutral	18.2%	23.3%	19.9%
Traditional Role	27.0%	30.8%	35.5%
Total	100.0%	100.0%	100.1%
N (total 2,544)	(955)	(884)	(705)

tau-*b* = .079 d_{rc} = .077 gamma = .123 lambda_{rc} = .000

SOURCE: Center for Political Studies, 1972 American National Election Study.

Table 8.13 Attitudes Toward Women's Role by Age, Controlling for Education

	Grade School Only			High School Only			Some College		
Role for Women	Under 35	35–55	Over 55	Under 35	35–55	Over 55	Under 35	35–55	Over 55
Equal	35.3%	34.9%	40.6%	47.3%	40.0%	46.8%	67.7%	61.2%	50.4%
Neutral	7.8%	18.6%	15.6%	22.4%	26.0%	20.9%	13.8%	20.9%	27.4%
Traditional	56.9%	46.5%	43.8%	30.2%	34.0%	32.3%	18.5%	18.0%	22.2%
Total	100.0%	100.0%	100.0%	99.9%	100.0%	100.0%	100.0%	100.1%	100.0%
N (total 2,541)	(51)	(129)	(288)	(526)	(477)	(297)	(378)	(278)	(117)
	tau-*b* = −.061			tau-*b* = .023			tau-*b* = .091		
	d_{rc} = −.065			d_{rc} = .022			d_{rc} = .036		
	gamma = −.106			gamma = .035			gamma = .156		
	lambda_{rc} = .000			lambda_{rc} = .000			lambda_{rc} = .000		

SOURCE: Center for Political Studies, 1972 American National Election Study.

the person tends to favor a traditional role for women or an equal role for women. Education is a particularly useful control when age effects are being studied. The younger generation often seems to have different attitudes than the older generation, but this may be at least partly due to their having more education than the older generation. An education control can reveal whether the supposed "generation gap" is really an "education gap." Table 8.13 restates Table 8.12, controlling for education. Interpret the results. Is there a generation gap in attitudes on this topic, or is it an education gap?

9

Changing Variables

Frequently, it is necessary to modify variables prior to an analysis. The variables in the data file may not be identical to those that you as analyst want to study. The concepts in the theory suggest certain types of operational indicators, which can be derived from (but are not identical to) the variables in the data file. In these cases, it is necessary to change the variables prior to generating the desired set of tables, correlations, or controls.

RECODING

Modifying a Variable

The first case of this type is when *variable recoding* is required. For example, the data may include a variable named "party identification" that shows to which political party the respondent feels closer. The variable may be coded:

1. strong Democrat
2. weak Democrat
3. Independent, leaning to Democrats
4. Independent, leaning to neither party
5. Independent, leaning to Republicans

6. weak Republican
7. strong Republican
8–9. missing data

But the theory being investigated may pertain to "strength of partisanship" rather than its direction. For example, the theory may be that the longer one has identified with a party, the stronger is that identification. The dependent variable for that relationship is strength of partisanship, without a direction component. Ideally, the strength of partisanship variable would be coded as follows:

1. strong identifier
2. weak identifier
3. independent leaner
4. pure independent
8. don't know
9. not ascertained

By a process of combining categories, the strength of partisanship variable can be generated from the original party identification variable that is available in the data:

new 1 = old 1 and old 7
new 2 = old 2 and old 6
new 3 = old 3 and old 5
new 4 = old 4
new 8 = old 8
new 9 = old 9

The new category 1 (strong identifier) includes those coded 1 (strong Democrat) and those coded 7 (strong Republican) on the original list. Similarly, the new category 2 (weak identifier) consists of those coded 2 (weak Democrat) and 6 (weak Republican) on the party variable. The other categories have been transformed in the same way. Conceptually, there is nothing complicated about this single-variable recoding, though it is important that the computer program used have an easy way of doing it.

Most statistical computer programs, such as SPSS, have the capability of transforming variables in this manner. The major programs differ in the way in which they are instructed to accomplish the recoding, but nearly all permit it. Once a variable is recoded, that new variable may be used in any tables or analyses (or other recodings) in which any of the original variables might have been used.

Combining Two Variables

A more difficult but nevertheless useful type of recoding is *bivariate recoding*—when two variables are recoded together to yield the variable that is of true interest. For example, one might have a theory that older voters vote against their party less often than younger ones (presumably because older voters are more confirmed in their partisanship than younger voters). An election study would have a variable showing the party the person identifies with (coded as in the previous example), a variable showing how the person voted (say, for president), but no variable showing whether or not they voted with their party. If we need this last variable for cross-tabulation with age, we would have to construct it from party identification and vote.

To do so, list the desired categories for the new variable, being sure to include all necessary missing data codes, such as:

1. vote with party
2. vote against party
9. no party, no vote, or missing data on either question

Say that the vote is coded:

1. Democrat
2. Republican
5. did not vote
8. don't know
9. missing data

The appropriate recoding would be:

new 1 (vote with party) = party identification 1, 2, 3 *and* vote 1 (Democrats voting Democratic) and
= party identification 7, 6, or 5 *and* vote 2 (Republicans voting Republican)

new 2 (vote against party) = party identification 7, 6, or 5 *and* vote 1 (Republicans voting Democratic) and
= party identification 1, 2, or 3 *and* vote 2 (Democrats voting Republican)

new 9 (missing data) = party identification 4, 8, or 9 (independents or missing data on party identification) and

= vote 5, 8, or 9 (did not vote or missing data on vote)

The result of the recoding is a new variable, which we might call "defection," showing whether or not the person defected from the party with which he or she identifies.

INDICES AND SCALES

Clearly the recoding process can be built up beyond two variables. Each new variable would be some combination of 1, 2, or 3, or even more variables which exist in the data file.

Additive Indices

A special case of multivariate recoding is *index construction*—in which several variables are combined to yield an index measuring the concept of interest. For example, a survey might ask people four separate questions concerning whether or not they obtained news about the election campaign from (1) television, (2) radio, (3) newspapers, and (4) magazines. For some purposes one would want to keep the differentiation among the four different media, seeing whether, for example, those who relay on television differ in their behavior from those who rely on newspapers. But for other purposes it would be useful to know from how many media the person obtains news. People who pay attention to several media could differ in their behavior from those who employ only a single news medium. Therefore we would want a count of how many media the person employs.

Say that each variable were coded:

1. employs this medium
0. does not employ this medium
9. missing data

We might make an index from the four separate media variables by just adding up people's scores on the four separate variables. A person with a score of "4" would employ all four media, "3" would mean the person employs any three media, and so on.

Missing Data. The missing-data category creates a problem here. It could be handled in several ways. One might give a missing-data code to anyone with missing data on *any* of the four questions. Or, more likely, we might decide not to count the 9's when adding the variables but to give a respondent a score of 9 if there is missing data for that respondent on *all* four questions. These are imperfect ways of dealing with missing data; unfortunately there are no perfect ways.

The final result would be a variable coded:

4. employs four media
3. employs three media (possibly missing data on the fourth)
2. employs two media (possibly missing data on others)
1. employs one medium (possibly missing data on others)
0. employs no media (possibly missing data on some)
9. missing data on all four media

This would be an *additive index* of the four media questions, with a minimum assignment of cases to missing data. (Remember when analyzing this variable, the "9" signifies missing data rather than nine media used.)

More Than One Dimension. Besides the problems with missing data, additive indices, such as in the above example, suffer from another major weakness: different variables might be tapping wholly different concepts. Consider, for example, category 2 on our media index. Respondents who employ only the print media—newspapers and magazines—would receive a score of 2 on the media index, as would respondents who employ only broadcast media—television and radio. These two types of people would have the same scores for very different forms of behavior. (It is not difficult to imagine that very different types of people employ only print media or only broadcast media. One would expect, for example, the broadcast-media people to have received less total information about a political campaign than if they had read detailed articles in the print media.) In other words, in this example the index may be measuring two different underlying concepts or "dimensions"—a broadcast-media dimension and a print-media dimension. Because there might be more than one dimension, we cannot expect all the respondents coded 2 on the index to be similar even if it would be advantageous to be able to do so.

Table 9.1 Media Use Scale

Medium Used	Valid Patterns A	B	C	D	E	Error Patterns W	X	Y	Z	Total
Television	yes	yes	yes	yes	no	no	no	no	yes	
Radio	yes	yes	yes	no	no	yes	no	yes	?	
Newspapers	yes	yes	no	no	no	yes	yes	no	yes	
Magazines	yes	no	no	no	no	yes	no	yes	yes	
Frequency	20	19	19	19	19	1	1	1	1	100
Score	4	3	2	1	0	4	0	?	4	
Number of errors	0	0	0	0	0	1	1	2	0	4
Number of yeses	4	3	2	1	0	3	1	2	?	
Second error count	0	0	0	0	0	2	2	2	?	6

Guttman's method: Number of errors = 4

Number of responses = (Number of people) × (Number of items) − Amount of missing data = (100) × (4) − 1 = 399

$$\text{Coefficient of reproducibility} = 1 - \frac{(\text{Number of errors})}{(\text{Number of responses})}$$

$$= 1 - \frac{4}{399} = .99$$

Minimum marginal reproducibility = .68

Coefficient of scalability = .97

Alternate method: 6 errors, 396 responses; Reproducibility = .98

NOTE: "?" is used to represent missing data.

Guttman Scaling

Clearly it would be of some value to know if an index were unidimensional (composed of only one dimension). Guttman scaling was developed to handle that need. Guttman scaling can be used to determine whether a set of variables measures a single concept or dimension and thus whether they can be combined. We will use our example above to illustrate how this would be done.

Presumably, television is the easiest medium to use, so most people would use it first; radio would be the next easiest, newspapers the next, and magazines the hardest. If media use were unidimensional, no one would employ a more difficult medium without first employing all of the easier ones. No one would use newspapers for campaign news unless he or she also used television and radio. Those who use magazines would use all of the other media. Thus, if media use were *perfectly* unidimensional, the only patterns that would appear would be:

4. uses all media
3. uses television, radio, newspapers
2. uses television, radio
1. uses television
0. uses no media
9. missing data

Table 9.1 illustrates the same notion in a slightly different form. Each column gives the responses of a set of respondents. The columns headed by the letters "A" through "E" correspond to the perfectly cumulative categories listed above. Below these columns are shown the five ordinal *scores* (4, 3, 2, 1, 0) the persons in each category would be given on a Guttman scale. If only the cumulative patterns "A" through "E" appeared in the data, then a perfect Guttman scale would be formed. The appearance of the other possible response patterns (such as those denoted "W" through "Z") is counted as an "error"—that is, as a divergence from a perfect Guttman scale.

Measuring the Extent of Cumulation. To determine the extent to which a set of questions fit a Guttman scale, it is necessary to count the number of errors that occur in the responses. Consider, for example, pattern "W." Since that pattern most closely matches valid pattern "A" (except that the person does not watch television), it would be given a score of 4 with one error. More generally, the number of errors for a pattern is the least number of responses that must be changed to obtain a valid pattern. "X" most closely resembles valid pattern "E," so it is scored 0 with one error. Sometimes an error pattern is equally close to more than one valid pattern. Thus, pattern "Y" could be scored 4 with two errors, 2 with two errors, or 0 with two errors. Scoring where data are missing is often possible, as with the pattern "Z," which is said to fit pattern A without errors. Note that altogether there are four errors present, given the frequencies of each response pattern in Table 9.1. The proportion of the total number of responses that fit the valid patterns is known as the *coefficient of reproducibility.* Its calculation is illustrated beneath Table 9.1. When the coefficient of reproducibility is a perfect 1.00, each person's responses are perfectly reproducible from the person's score on the scale.

Guttman originally suggested that a reproducibility value of at least .90 should be required for a good scale. However, experience with that criterion suggests that it is too low and that a .95 criterion might be more realistic. Additionally, a *coefficient of scalability* has been developed to show how much better the reproducibility is than the minimum reproducibility that might be expected given the marginal frequencies of the responses to each question in the scale. The point behind this measure is that if, say, 90 percent of the respondents give the same answer on a question, then at most 10 percent of the responses on that question can produce scale errors (since there would be no error if all the responses were identical). Consequently, we can find how low the coefficient of reproducibility could possibly be (given the numbers of respondents that answered the questions in each way) and then compute the improvement of the actual reproducibility over that minimum value. The coefficient of scalability is the ratio of the actual improvement to the maximum potential improvement. For example, while the coefficient of reproducibility for the scale in Table 9.1 is .99, it could not possibly be lower than .68, given the marginal distributions of the answers. The actual improvement in reproducibility is .99 − .68 = .31, while the maximum potential improvement is 1.00 − .68 = .32. Thus the actual improvement is 97 percent (.31/.32) of the potential improvement, and the coefficient of scalability has a value of .97. Generally the coefficient of scalability should be at least .60 for a good scale.

Because the scoring and error-counting procedures outlined so far are computationally rather difficult and time consuming, a simpler and quicker procedure is often substituted. The greater speed of this second procedure results in its being used in most computer programs for Guttman scaling (including the SPSS Guttman scaling procedure). While both procedures are usually referred to as Guttman scaling, their different methods do give slightly different results. For example, pattern "W" would receive a score of 3 for the fast procedure because the respondent uses three media; pattern "W" differs from the valid pattern with three media used ("B") by using magazines (one error) but not television (a second error). Thus pattern "W" would be given a score of 1 (since one medium is used) but with two errors, and pattern "Y" would be given a score of 2 with two errors. Also, in order to speed calculation, patterns containing missing data are dropped from the analysis in this second procedure. This rapid Guttman-scaling procedure does produce some scores that differ from those produced by classical Guttman scaling, but the two sets of scores would be highly correlated with each other (and would have similar correlations with other variables). The quick procedure also yields an error count that is about twice as high as the classic Guttman count, which immediately lowers the reproducibility. Under this error-counting procedure, a .90 reproducibility would be considered very good.

We should mention that another test for the cumulation of the variables would be to use the gamma statistic, described in Chapter 7. to measure the cumulation between each pair of variables. Indeed, this is the best use of the gamma statistic. If the gamma computed between each pair of variables is at least .80, then the items together form a respectable Guttman scale.

Number of Dimensions. If the variables are not cumulative, then they do not measure the same underlying concept or dimension and should not be combined into a Guttman scale (or even an additive index). Sometimes it makes sense to drop a variable that is not cumulative with the remaining variables. Other times it makes sense to construct two separate dimensions, such as one for print media and one for broadcast media in our example.

It should be remembered that whether a set of questions can be formed into a Guttman scale is an empirical question. A scale with a sufficiently high coefficient of reproducibility may or may not exist. In fact the mere determination of the existence of such a scale may be the goal of a researcher. In such a circumstance the researcher might argue that a set of questions on governmental policy all represent the same dimension. For example, the researcher might argue that a set of policies are all part of the liberal-conservative dimension because they form a Guttman scale. In other situations, however, the researcher forms the scale as the first step in a more complex scheme of analysis. For example, individuals could be scored on a liberal-conservative scale so that these scores can be related to income and education characteristics in a series of tables.

Often, it becomes impossible to account for a set of attitudes with a single Guttman scale. For example, if we had a set of questions on international and domestic policy, we might find that those who were liberal on domestic issues are not necessarily those who are liberal on international issues. Separate Guttman scales would then be needed for domestic and international questions. We would say that there are two dimensions underlying the data. In other circumstances, more than two dimensions may be required.

Multidimensional Scaling

We have mentioned the use of a correlation coefficient, gamma, to record the extent of cumulation between a pair of variables. More generally, the correlation coefficients (including the tau's) can be used to learn something of the structure among a set of attitudes. The more two variables measure the same thing, the greater would be their correlation coefficient—and vice versa, the greater the correlation coefficient for a pair of variables, the more similar are the two variables.

Spatial Representation. A geometric space can be used to represent the similarities of the variables. The space would use as few dimensions as possible, consistent with the desire to represent the data accurately. A space of one dimension might suffice, but the relationships between the variables might be so complicated that a space of several dimensions (a "multidimensional" space) might be necessary. The space could even consist of more than three dimensions, in which case it could not be portrayed graphically.

In accord with the above logic, the greater the correlation between a pair of variables (and the more similar they are), the closer together we would want to put them in the space. If we had several variables, this would result in similar variables being placed near one another and dissimilar ones being placed far apart in the space. Following this approach, "multidimensional scaling" procedures have been developed to locate variables in a space such that the rank order of distances between each pair of variables is the opposite of the rank order of the correlation coefficients. (The pair of variables furthest apart are the pair with the lowest correlation, and so on.) Although these procedures are theoretically very simple, they are very difficult and time consuming to calculate; hence, they are always done on a computer.

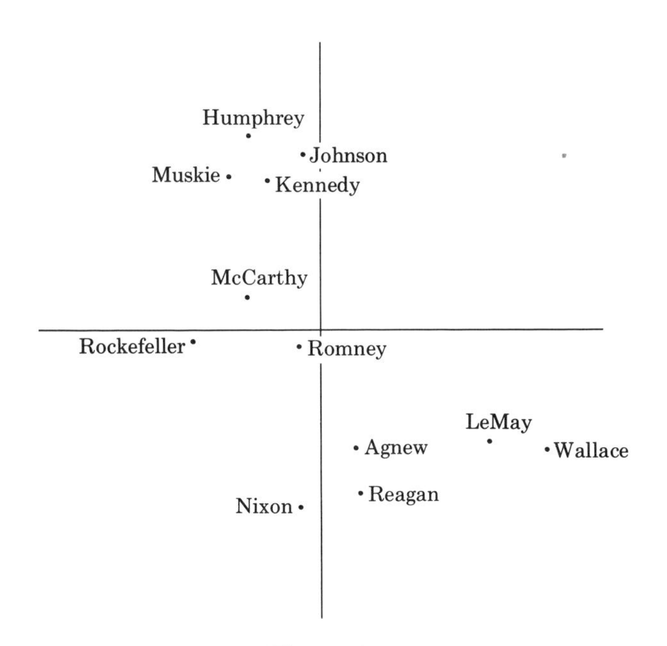

Figure 9.1
The 1968 Candidate Space in Two Dimensions
(From Herbert F. Weisberg and Jerrold G. Rusk, "Dimensions of Candidate Evaluation," *American Political Science Review, 64,* 1970, p. 1176. Reprinted by permission of the publisher.)

An Example. Figure 9.1 illustrates the output of a computer program for this type of multidimensional scaling. The data analyzed are the reactions to presidential contenders in 1968 on the part of a national cross-section sample. People were asked to rate the candidates on a feeling thermometer—a score of "100" for Kennedy showed the respondent felt very warm toward Kennedy, and a score of "0" showed that the person felt very cold toward him. The reactions to each pair of candidates were correlated. If people who liked Kennedy also invariably liked Humphrey, the correlation between the ratings of Kennedy and Humphrey might be near $+1.0$. In general, a positive correlation between two candidates means that the more one candidate is liked, the more the other is. By contrast, a negative correlation means that the more one candidate is liked, the less is the other. For example, the more people liked Humphrey in 1968, the less they liked his opponent, Nixon.

Multidimensional scaling of the Pearson r correlations is shown in Figure 9.1. As much as possible in a two-dimensional representation, this figure is drawn so that greater correlations correspond to smaller distances. Kruskal's Multidimensional Scaling computer program has been used to scale the correlations shown here. Note the obvious cluster of mainstream Democrats at one end of the space, the mainstream Republicans opposite them at the end of one dimension, Wallace and his running mate opposite the Democrats at the end of another dimension, and the less well known and less partisan set of moderates (Rockefeller, Romney, and McCarthy) in the middle of the space.

We could try to name these dimensions. The authors of the study which obtained Figure 9.1 found the vertical dimension to be associated with "partisan issues" and the horizontal dimension to be associated with the "social issues" of the 1960's, such as Vietnam and urban unrest. Naming dimensions can be difficult, and different researchers frequently interpret the same dimension in different ways. Consequently, some analysts examine the structure of attitudes without pinning names on the dimensions.

SUMMARY

We often require different forms of variables than are available in the data set. Simple recoding often suffices to construct the needed variables. When several variables are recoded into an index, it makes sense to check first whether they all tap the same basic dimension. Guttman scaling is one test of such unidimensionality. Guttman scaling also turns our attention to the structure of a set of variables. The correlations between variables convey information about the structure of the underlying attitudes. One means of portraying this structure is multidimensional scaling, which uses spatial distances to represent similarity between variables. In the next chapter we turn to the techniques used for analysis of interval data, including an important alternative to multidimensional scaling for interval level data.

Questions

How would you handle the following situations?

1. You want to determine whether younger people are more likely to consider themselves political independents than are older people. However, the

only available measure of partisanship is coded: (0) strong Democrat, (1) weak Democrat, (2) independent leaning to Democrat, (3) pure independent, (4) independent leaning to Republican, (5) weak Republican, (6) strong Republican. Construct a new variable coded: (1) independent (including independent leaners), (2) partisan.

2. You want to determine whether strength of partisanship is related to whether a person votes a split ticket (votes Republican for some offices and Democratic for others). However, the only available measures of voting are presidential vote, coded: (1) Democrat, (2) Republican, (3) not vote; and congressional vote, coded: (1) Democrat, (2) Republican, (3) not vote. Construct a new variable coded: (1) straight ticket, (2) split ticket, (3) skipped voting for at least one race.
3. You hypothesize that a person's education affects his or her amount of campaign participation. However, the available participation questions are separate questions on attendance at political meetings coded: (1) yes, (0) no, (9) missing data; working for a party or candidate coded: (1) yes, (0) no, (9) missing data; and campaign contributions coded: (1) yes, (0) no, (9) missing data. How would you construct an additive index from these questions to yield an overall measure of campaign participation?
4. If the above campaign-participation questions formed a Guttman scale, with party meetings being the most frequent activity and campaign contributions being the least frequent activity, then what response patterns would fit the scale perfectly?

10

Statistical Inference

When using small samples, it is important to be sure that findings are not just due to chance. If the sample is small, the odds are great that it is atypical. Another sample the same size might give very different results. Consequently, one does not want to overgeneralize on the basis of small samples. Procedures are required to indicate what inferences to the target population can be made safely. Therefore, in this chapter we shall introduce the statistical-inference procedures that are used for small simple random samples. In order to do this, we shall first have to present notions from probability theory.

PROBABILITY THEORY

We often speak of the "probability" of an event, such as the probability of drawing a red card from an ordinary deck of playing cards. A probability is a number between 0 and 1 that is associated with an event. A common interpretation of probabilities (although not the only possible interpretation) relates to the relative frequency of events. For example, the relative frequency of red cards in a card deck is the ratio of the number of red cards (26) to the total number of cards in the deck (52), $26/52 = .50$; so the probability of drawing a red card from a full deck is .50. By a similar logic, the probability of drawing a black card from the deck would also be .50, the probability of drawing a heart would be .25, the probability of drawing a diamond would be .25, the probability of drawing an ace would be $4/52 = .08$, and so on.

If we draw one card from the deck, replace it, reshuffle, draw again, and keep doing this sampling with replacement a large number of times, the proportion of times that we draw a red card will be very close to .50. Thus, probabilities refer not only to the relative frequency of events but also to the long-term rates of occurrence.

Two events are described as "mutually exclusive" if both cannot occur at once. For example, the events of drawing a red card and drawing a black card are mutually exclusive, because when one card is drawn it cannot be both red and black. Probabilities of mutually exclusive events add up, so the probability of drawing a red card or a black card is $.50 + .50 = 1.00$, or certainty. Similarly, drawing a heart and a diamond are mutually exclusive events, so the probability of drawing either a heart or a diamond is $.25 + .25 = .50$, the probability of drawing a red card. Probabilities of non-mutually exclusive events do not sum in this manner. For example, the events of drawing a heart and of drawing an ace are not mutually exclusive (because drawing an ace of hearts would satisfy both events at once), and so the probability of drawing either a heart or an ace (16/52 is the proportion of cards that are either hearts or aces or both) does not equal the sum of their separate probabilities ($13/52 + 4/52 = 17/52$).

A set of events is called "exhaustive" if all possible outcomes are accounted for by the events. Drawing a red card and drawing a black card are exhaustive events, because any card drawn from the deck satisfies one event or the other.

We can now finally define the key concept of a "probability distribution." A probability distribution states the probabilities of a set of mutually exclusive and exhaustive events. For example, the probability distribution associated with the color of a card in an ordinary deck of playing cards is .50 for a red card and .50 for a black card. Similarly, the probability distribution for the suit of a card in an ordinary deck of playing cards is .25 for a heart, .25 for a diamond, .25 for a club, and .25 for a spade. Note that the probabilities sum to 1.00 in each case, because the events are mutually exclusive and exhaustive.

The Normal Distribution

Some theoretical distributions are particularly important in statistics. The most important one is the so-called normal distribution. The shape of the distribution follows a precise mathematical equation

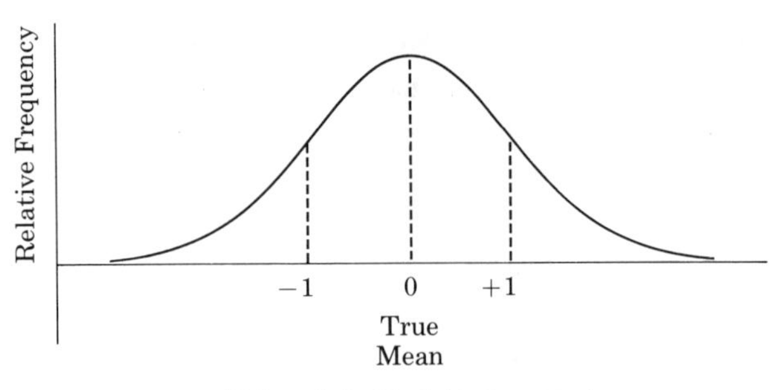

Figure 10.1
The Normal Distribution

that is graphed in Figure 10.1. The horizontal axis shows the value of the variable, and the vertical axis is the relative frequency of each value.

The horizontal axis of Figure 10.1 is calibrated in a special way—it is measured in standard deviation units from the mean, called "z scores." Recall from Chapter 5 that the standard deviation is a measure of the dispersion of the values of the variable. Z scores are computed according to the following formula:

$$z = (X - \overline{X})/s$$

where $\overline{X}$ is the mean value of the variable X and s is its standard deviation. For example, if the mean of a variable is 10 and its standard deviation is 5, then a score of 15 would be one standard deviation unit above the mean ($z = +1$), a score of 20 would be two standard deviation units above the mean ($z = +2$), a score of 5 would be one standard deviation unit below the mean ($z = -1$), and so on. A z score of 0 always represents the mean of the variable.

The center of the normal curve is the true mean of the variable. Values of the variable closest to the true mean occur most frequently, while values farthest away occur least frequently.

Areas Under the Normal Curve

The normal curve is constructed such that the area between it and the horizontal axis is one unit. The area under the curve is interpreted as a probability. For example, the probability of a value of a variable falling between the lowest possible value on the horizontal axis, and the highest possible value on the horizontal axis is the area under the entire curve or 1.0, certainty.

Because the true mean is at the center of the distribution, the probability of a value being higher than the true mean would be equal to the area under the curve to the right of the true mean, or half the total area under the curve—.50. The curve is symmetric. Therefore, the probability of a value being below or lower than the true mean is exactly the same as the probability of its being above—.50.[1]

The probabilities of other values of the variable can also be obtained from the normal curve. For example, the probability of obtaining a value more than one standard deviation below the mean is .16; that is, the area under the curve to the left of one standard deviation unit is .16. Because of the symmetry, the probability of obtaining a value more than one standard deviation above the mean is also .16. Thus, the probability of obtaining a value no greater than one standard deviation unit either side of the mean is $(1.0 - .16) - .16 = .68$.

To be more concrete, say that scores on the Graduate Record Examination (GRE) are normally distributed with a mean of 500 and a standard deviation of 100. Applying the principle stated in the previous paragraph, the probability of a score under 400 (the mean of 500 minus one standard deviation) would be .16, the probability of a score above 600 (500 + 100) would be .16, and the probability of a score between 400 and 600 (plus or minus one standard deviation unit around the mean) would be .68. These values are shown in Figure 10.2. It is important to see what this is saying and what it is not saying. It does not mean that you have a .16 chance of scoring below 400 on the GRE. Instead it means that 16 percent of the students taking the exam score below 400, so the probability that a randomly drawn individual has a score below 400 is .16.

Note that the values in the tails (parts farthest from the center) of the distribution are particularly unlikely. Extreme values are much less likely than moderate values. In fact, the curve shows that the

---

[1]What about the probability of exactly obtaining the mean value? The normal curve is used with continuous numerical variables, such as length. There is virtually zero probability of a pencil being exactly 3 inches long. It might be 3.00012358 . . . inches, but the chances of it being any particular value is infinitesmal. Therefore, we can disregard the possibility of the variable having a value exactly equal to the mean. More generally, the normal curve shows the probability of values falling into particular ranges, not the probability of specific values.

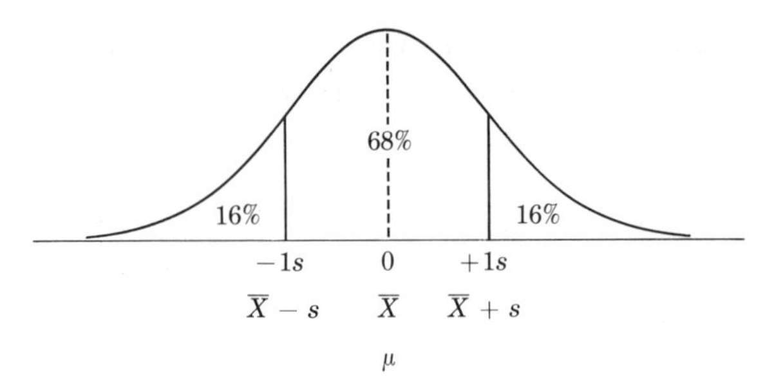

Figure 10.2
One Standard Deviation Unit Range for the Normal Distribution

probability of obtaining any range of values declines as the range moves away from the mean either below (to the left) or above (to the right). The probability of obtaining a value more than 1.96 standard deviation units below the mean is only .025. By symmetry, the probability of obtaining a value more than 1.96 standard deviations above the mean is also .025. Therefore, the probability of obtaining a value not more than 1.96 standard deviation units from the mean (in either direction) is $(1.0 - .025) - .025 = .95$ (see Figure 10.3). In the GRE example above, this corresponds to a .95 probability of obtaining a score between 304, or $500 - (1.96 \times 100)$, and 696, or $500 + (1.96 \times 100)$. The chance of obtaining extreme scores below 304 and above 696 is only .05.

Since it is difficult to read the area from a curve such as shown in Figure 10.1, statisticians have constructed tables from which the area can be read more easily. Table 10.1 gives the probabilities of particular ranges of values under the normal curve. To read the table, you would decide the number of standard deviation units (z) above the mean that interests you. The table shows what proportion of the area under the normal curve is *above* that z value. For example, to find the area above the mean itself, look under .00 standard deviations above the mean, and you will find in row .0 and column .00 the number .5000, indicating that half the area is above the mean. To find the area above one standard deviation unit, look under z of 1.00—row 1.0 and column .00—and you will find the number .1587, which we rounded to .16 for Figure 10.2 and in the above example. To find the area above .45 standard deviation units above the mean, use row .4 and column .05 to find the value .3264.

The normal-curve table can also be used to obtain the z values (number of standard deviation units above the mean) associated with certain areas. To find the z value for which there is only a .025 probability of higher values, find .025 in the body of the table—it is in row 1.9 and column .06, so the z value is 1.96. Similarly, to find the z value for which there is only a .005 probability of higher values, find .005 in the body of the table—it is in row 2.5 and between column .07 and .08, so the z value is about 2.575. (More precice tables indicate it is actually 2.576.)

The table only gives probabilities for positive z's. However, the symmetry of the normal curve permits the same table to be used for negative z's. If you want to know the proportion of the area under the curve less than $z = -1.28$, you would look under row 1.2 and column .08 to get the value .1003.

The Importance of the Normal Distribution

The normal curve is important for a number of reasons. Some variables have normal distributions, although most social-science variables do not have normal distributions. Minimally, we may assume that the errors in measurements for a variable have a normal distribution. That assumption considerably simplifies the mathematics in some statistical proofs. The normal distribution often serves as a good approximation to other distributions, such as the number of heads in 50 tosses of a fair (unbiased) coin. Finally, and most important, even when a variable does not have a normal distribution, the mean of that variable for a large sample can be regarded as having

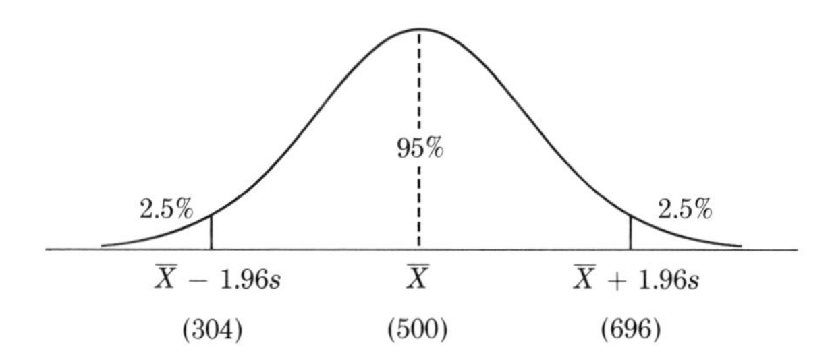

Figure 10.3
Ninety-Five Percent Range for the Normal Distribution

Table 10.1 Areas Under the Normal Curve. (Entries show the probability of obtaining a z value above z_0. Areas are found by symmetry for negative values of z_0.)

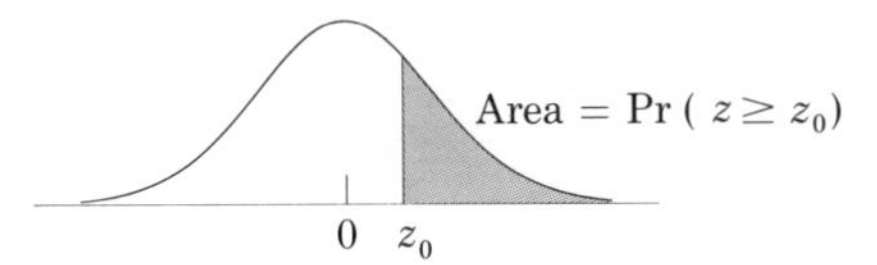

	Second Decimal Place of z_0									
z_0	.00	.01	.02	.03	.04	.05	.06	.07	.08	.09
0.0	.5000	.4960	.4920	.4880	.4840	.4801	.4761	.4721	.4681	.4641
0.1	.4602	.4562	.4522	.4483	.4443	.4404	.4364	.4325	.4286	.4247
0.2	.4207	.4168	.4129	.4090	.4052	.4013	.3974	.3936	.3897	.3859
0.3	.3821	.3783	.3745	.3707	.3669	.3632	.3594	.3557	.3520	.3483
0.4	.3446	.3409	.3372	.3336	.3300	.3264	.3228	.3192	.3156	.3121
0.5	.3085	.3050	.3015	.2981	.2946	.2912	.2877	.2843	.2810	.2776
0.6	.2743	.2709	.2676	.2643	.2611	.2578	.2546	.2514	.2483	.2451
0.7	.2420	.2389	.2358	.2327	.2296	.2266	.2236	.2206	.2177	.2148
0.8	.2119	.2090	.2061	.2033	.2005	.1977	.1949	.1922	.1894	.1867
0.9	.1841	.1814	.1788	.1762	.1736	.1711	.1685	.1660	.1635	.1611
1.0	.1587	.1562	.1539	.1515	.1492	.1469	.1446	.1423	.1401	.1379
1.1	.1357	.1335	.1314	.1292	.1271	.1251	.1230	.1210	.1190	.1170
1.2	.1151	.1131	.1112	.1093	.1075	.1056	.1038	.1020	.1003	.0985
1.3	.0968	.0951	.0934	.0918	.0901	.0885	.0869	.0853	.0838	.0823
1.4	.0808	.0793	.0778	.0764	.0749	.0735	.0722	.0708	.0694	.0681
1.5	.0668	.0655	.0643	.0630	.0618	.0606	.0594	.0582	.0571	.0559
1.6	.0548	.0537	.0526	.0516	.0505	.0495	.0485	.0475	.0465	.0455
1.7	.0446	.0436	.0427	.0418	.0409	.0401	.0392	.0384	.0375	.0367
1.8	.0359	.0352	.0344	.0336	.0329	.0322	.0314	.0307	.0301	.0294
1.9	.0287	.0281	.0274	.0268	.0262	.0256	.0250	.0244	.0239	.0233
2.0	.0228	.0222	.0217	.0212	.0207	.0202	.0197	.0192	.0188	.0183
2.1	.0179	.0174	.0170	.0166	.0162	.0158	.0154	.0150	.0146	.0143
2.2	.0139	.0136	.0132	.0129	.0125	.0122	.0119	.0116	.0113	.0110
2.3	.0107	.0104	.0102	.0099	.0096	.0094	.0091	.0089	.0087	.0084
2.4	.0082	.0080	.0078	.0075	.0073	.0071	.0069	.0068	.0066	.0064
2.5	.0062	.0060	.0059	.0057	.0055	.0054	.0052	.0051	.0049	.0048
2.6	.0047	.0045	.0044	.0043	.0041	.0040	.0039	.0038	.0037	.0036
2.7	.0035	.0034	.0033	.0032	.0031	.0030	.0029	.0028	.0027	.0026
2.8	.0026	.0025	.0024	.0023	.0023	.0022	.0021	.0021	.0020	.0019
2.9	.0019	.0018	.0017	.0017	.0016	.0016	.0015	.0015	.0014	.0014

SOURCE: Thomas H. Wonnacott and Ronald J. Wonnacott, *Introductory Statistics*, 2nd. ed. (New York: Wiley, 1972), p. 480. Reprinted by permission of John Wiley & Sons, Inc. After R. E. Walpole, *Introduction to Statistics*, Macmillan.

come from a normal distribution. This result is known as the "central-limit theorem" and is important enough that we will explain it in some detail.

THE CENTRAL-LIMIT THEOREM

Sampling Distribution

We first need some additional terminology. There is a distribution of cases on a variable. When we take a random sample, there is a distribution of sample values on the variable—the sample distribution. We can calculate the mean for that sample. Say we take a second independent random sample and calculate its mean, we take a third and calculate its mean and so on for a large number of samples. If we look at all of our sample means together, we have a distribution of sample means, known as the "sampling distribution of means," or often simply as the "sampling distribution." According to the central-limit theorem, for very large numbers of large samples, the sampling distribution of means is approximately normal; that is, it is approximately the same shape as the normal distribution.

As an example, assume GRE scores have a true mean of 500 and a standard deviation of 100 for the population. We give this test to one sample of people and get a mean of 510 for that sample. Another sample might have a mean of 485, and so on. If our samples are large and we take a lot of them, these means would have a normal distribution. That distribution of sample means is what we are calling the *sampling distribution.*

Standard Error

The mean of the sampling distribution is the same as the population mean of the variable. The standard deviation of the sampling distribution is what we termed in Chapter 5 the *standard error of the mean.*

$$s_{\mathrm{m}} = \sqrt{s^2/(N-1)} = s/\sqrt{N-1}$$

If we have a sample of size 26 for the GRE example above, the standard error would be 20 (the standard deviation of 100 divided by the square root of $26 - 1$).

We now know that GRE sample means have a normal distribution with a mean 500 and a standard error of 20. Recall our earlier results on the probability of values with a normal distribution. There is a .68 probability of a value being within one standard deviation of the mean; hence there is a .68 probability of the sample mean being between 480 and 520 for samples of size 26. There is only a .05 probability of a value being farther than 1.96 standard deviations away from the mean, so there is only a .025 probability of the sample mean being below 460.8, or $500 - (1.96 \times 20)$ and a .025 probability of it being above 539.2, or $500 + (1.96 \times 20)$.

You may be wondering why we are using 20 for our calculation of the range of sample values for the mean in this last example but used 100 in the earlier example. In the earlier example, we were examining the distribution of cases so we used the standard deviation (100). When we examined the distribution of possible means in the more recent example, we used the standard error of the mean (20), an estimate of the standard deviation of the sampling distribution.

Note how the standard error operates. The larger the sample, the smaller the standard error. As a result, the bounds on the sample means will be narrower for larger samples. If the sample in the GRE example were of size 101, the standard error would be 10. There would be a .68 probability of the sample mean being between 490 and 510 and only a .05 probability of the sample mean being either below 480.4 or above 519.6. (Note that the sample had to be nearly four times as large to cut the standard error in half.) This is what we should expect: There should be less chance of a large sample giving an atypical mean (here, a mean far away from the population mean of 500) than of a small sample giving an atypical mean.

Actually, there is an important statistical result, called the *law of large numbers,* which makes this point very generally. For random samples, the larger the sample size, the more likely it is that the sample mean is very close to the population mean. A small sample can have an atypical mean, but it is less likely that a large sample will. If the sample is large enough to include the entire population, then there will be no difference at all between the sample mean and the population mean.

While we have noted the standard error's dependence on the sample size, clearly, the standard error of the mean also depends on the standard deviation. The more dispersed the original observations, the less confidence there will be that the observed mean is the correct one. Conversely, if there is a little dispersion in the original observations, one can have more confidence in a sample of the same size.

Assumptions

The central-limit theorem makes no assumptions about the shape of the variable's distribution. Even if a variable has a distribution that looks very different from the normal curve shape (for example, even if a variable is bimodal so it takes on only very small or very large values but never moderate values), with large samples the sample mean will be very close to the population mean and the sample mean's distribution will approximate a normal distribution. Thus, it is the central-limit theorem that makes the normal distribution so important for data analysis.

The central-limit theorem requires that the population variance of the variable be known. Rarely are we in a position to know the true population variance for a variable. For large samples, this is no problem—the normal distribution still holds even when we must estimate the population variance with the sample variance. For small samples, the sampling distribution of means has a shape that is near that of the normal distribution but that is not exactly the same. The distribution of sample means for small samples has the shape of what we call the "*t*-distribution." We shall present this distribution in greater detail at the end of this chapter. Statistical inference with the *t*-distribution is very similar to statistical inference with the normal distribution except that there will be a slightly greater chance for extreme means.

So far we have made one other assumption that is unusual: that the population mean is known. We have looked for the probability of an atypical sample mean given a known population mean. In practice, the population mean is not known, and we do not take repeated samples in order to get a sampling distribution of means. Usually, all we have is a single sample mean and its standard deviation. In statistical inference, we reverse the central-limit theorem. We act as if our one sample mean were a sample from a normally distributed sampling distribution of means. The central-limit theorem indicates that if the population mean were known, a sample mean far away from it would be unlikely. Even if the population mean were not known, it should be rare for the population mean to be far away from the sample mean. Therefore, we will be able to make inferences about the population mean on the basis of the sample mean. There are actually several different approaches for statistical inference. We begin by describing the most classical: hypothesis testing.

HYPOTHESIS TESTING

The Logic of Null Hypotheses

Many social scientists feel that the first step in properly conducted research is to develop hypotheses about the variables being studied. Hypotheses are propositions that are deduced logically or at least derived from theory; in the case of political science, hypotheses would be based on theories of the behavior of political actors. The hypothesis posits something about a variable, such as its mean value or its relationship to another variable. For example, our theories of voting behavior might suggest the research hypothesis that Democrats are more likely than Republicans to vote for Democratic candidates.

Science is concerned with testing hypotheses. However, the rules of logic intrude here in a particular way. It is necessary to avoid committing what logicians term "the fallacy of affirming the consequent." Consider the claim "Condition A causes condition B." Observing that condition B has occurred might then seem to prove that condition A is true. However, this argument is fallacious, because it might be that condition C also causes condition B, and observing that condition B has occurred cannot by itself indicate whether condition A or C is true. For example, consider the claim "Cities with heavy rain have high annual precipitation counts." Discovering that a city has a high precipitation count, however, does not prove that it has had heavy rain. After all, heavy snow also leads to high precipitation counts. Observing a high precipitation count cannot by itself indicate whether heavy rain or heavy snow is the cause.

Returning to the voting example, finding that Democrats voted for the Democratic candidate in a particular election more than the

Republicans did cannot prove the generalization that Democrats are more likely than Republicans to vote for Democratic candidates. The claim studied would be "Democrats are more likely than Republicans to vote for Democratic candidates, so Democrats voted in a greater proportion for the Democratic candidate in the particular election studied than did Republicans." Looking back at the form of the fallacy of affirming the consequent, it should be clear that verifying that the consequent is true (that Democrats did vote in a greater proportion for the Democratic candidate in the particular election than did Republicans) cannot be used to prove that the research hypothesis (Democrats are more likely than Republicans to vote for Democratic candidates) is correct. The research hypothesis states a generalization, and we have data only on particular consequences of that generalization. We cannot prove that the hypothesis is true based on specific data.

The rules of logic do permit a different type of proof. "Denying the consequent" is a valid form of argument. Consider again the claim "Condition A causes condition B." Observing that condition B has *not* occurred then shows that condition A has *not* occurred. If "cities with heavy rain have high annual precipitation counts," then a low precipitation count for a city proves that it has not had heavy rain. Similarly, finding that Republicans voted in a greater proportion for the Democratic candidate in the particular election studied than did Democrats would show that the claim that "Democrats are more likely than Republicans to vote for Democratic candidates" is false. The rules of logic are different for proof and disproof, and as a result science proceeds by a series of disproofs rather than by direct proofs.

Because of the logical problem of proving a research hypothesis, scientists have developed the notion of the null hypothesis (H_0). The null hypothesis is usually the opposite of the research hypothesis (H_1). In our example, the null hypothesis would be "Republicans are at least as likely as Democrats to vote for Democratic candidates." We can use data to disprove this hypothesis without committing any logical errors.

Remember, however, that even if we have disproved or rejected our null hypothesis, we still have not proven the research hypothesis. At best, by offering evidence that the null hypothesis is not true, we only show that the research hypothesis *may* be true. In spite of this warning, many researchers speak of "accepting" the research hypothesis. We are no exception, but still this does not mean that by accepting the research hypothesis one has proved anything, in the logical sense.

Because disproving a null hypothesis is a valid form of argument, scientific research is often framed in terms of null hypotheses. Frequently the null hypothesis is that there is no relationship between two variables or that one variable does not affect another variable, but null hypotheses can be more general than this. In any case, the researcher is faced with the question of whether the null hypothesis should be rejected.

To complicate matters further, the hypotheses generally deal with relationships for the population of interest, while the data are only from a sample. As we are about to see, statistical inference is the process of determining whether sample data cause rejection of a null hypothesis about a population.

Types of Error

The researcher can either reject or not reject the null hypothesis. Obviously, if it is false, the researcher wants to reject it, while she does not want to reject it if it is true. Two types of error can be made in this situation (see Table 10.2). First, we could accidentally reject the null hypothesis when it is true. This type of error, falsely rejecting a true null hypothesis, is known as Type I error. Second, we might accept a null hypothesis when it is false. Accepting a false null hypothesis is known as Type II error.

We would like to avoid both types of error. However, the two types of error are related. The more stringent the criterion for rejection of the null hypothesis, the fewer true null hypotheses we will mistakenly reject (less Type I error), and the more false null hypotheses we will keep by accident (more Type II error). It is the same criterion that protects us from wrongly rejecting a true null hypothesis (wrongly accepting our research hypothesis) that causes us to err more often in accepting a false null hypothesis. Thus there is a tradeoff between these two types of error.

Scientists generally view Type I error as the more serious type. In medicine, one would not want to change treatments if the new treatment were not a real improvement over the safe, established treatment, so one would not want to take a large chance of falsely rejecting a true null hypothesis of no difference between the treatments. Only a slight chance of Type I error should be permitted. Merely continuing with the old treatment when the new one is better

Table 10.2 Error Conditions When Accepting or Rejecting a Null Hypothesis, Depending on Whether the Null Hypothesis Is True or False

	H_0 true	H_0 false
accept H_0	Correct	Type II Error
reject H_0	Type I Error	Correct

is preferred to wrongly changing treatments, so Type II error may be ignored.

In social science, the null hypothesis might be that the hypothesis of a previous researcher reported in the journals was correct. New research, on the other hand, may suggest a different relationship—the research hypothesis. The new research could be wrong, in which case publishing it would constitute rejecting a true null hypothesis (Type I error). Or the new research might be right, in which case not publishing it would constitute accepting a false null hypothesis (Type II error). The usual conservative statistical procedure is to take only a small chance of making a Type I error, even if doing so entails a large chance of Type II error. Therefore, hypothesis testing is set up so that one takes only a small chance of rejecting a true null hypothesis, even if this means that one often does not reject a false null hypothesis and that a new true research hypothesis is not accepted.

In hypothesis testing, the researcher actually sets the chance she is willing to take of making a Type I error. A common level in social science is the .05 level. That is, the test is set up so that there is only a .05 probability of rejecting a true null hypothesis. (The likelihood of a Type II error is usually ignored in this procedure.) Some researchers employ more stringent levels, such as the .01 level.

Testing Procedures

As an example of hypothesis testing, assume that we know that scores on the GRE have a normal distribution and a standard deviation of 100. We take a sample of size 26 from your university and obtain a sample mean of 541. We know that the national average on the test is

500. Does the sample show that your school differs from the national value? The null hypothesis is that the population mean for your school is 500. With such a small sample, we should be cautious in concluding the null hypothesis is wrong. The hypothesis-testing procedure is therefore useful. As shown in the last section, the standard error of the mean for this example is 20. Thus, our sample mean is more than two standard error units away from the null-hypothesis mean. If the null hypothesis were correct, the probability of a sample mean more than two standard error units away would be less than .05. So we can reject the null hypothesis at the .05 level. Your school's mean is significantly different from the national mean of 500.

Critical Ratio

At this stage, we should switch to a more mathematical presentation. What actually would be done in hypothesis testing would be to compute a test statistic known as the *critical ratio,* or z. The critical ratio is the difference between the sample mean ($\overline{X}$) and the null-hypothesis mean (which we denote by the Greek letter *mu,* μ), divided by the standard error of the mean: $z = (\overline{X} - \mu)/s_{\mathrm{m}}$. The normal curve table gives the z values required for different levels of significance. ("Statistical significance" at a particular level means that the probability of rejecting a true null hypothesis is less than or equal to that level.) The z value required for significance at the .05 level is 1.96. In our example, the z value is $(541 - 500)/20 = 2.05$. Since it is greater than 1.96, we would reject the null hypothesis. If z (the critical ratio) were less than 1.96, we would not reject the null hypothesis.

Critical Region

An equivalent procedure for hypothesis testing involves setting up a "critical region." It is a region of values such that a sample mean in the region would lead to rejection of the null hypothesis. For a normal distribution, there would be a 5 percent (or .05) chance of obtaining a value more than 1.96 standard errors from the population mean. When the population mean for the GRE is 500 (the null hypothesis) and the standard error 20, there is a .05 chance of obtaining a sample-mean GRE below 460.8, or 500 − (1.96 × 20), or above 539.2, or 500 + (1.96 × 20). Therefore, the critical region in which we would reject the null hypothesis is below 460.8 and above 539.2. The critical region corresponds to the darkened area in Figure 10.4. In our example above, the sample-mean GRE was 541; hence we would reject the null hypothesis.

The advantage of this formulation is that it immediately indicates what action we would take for any sample mean. We would not reject the null hypothesis if the sample-mean GRE were 535; if the sample-mean GRE were 455, we would reject the null hypothesis; and so on. Thus, the critical-region calculation provides all the information of the critical-ratio test but is more useful.

Probability Value

The critical ratio and the critical region share a disadvantage: They are so tied to a particular level of significance (the .05 level in our example) that another researcher who wanted to use a different level of significance would be unable to make use of the published result. The alternative is to calculate the probability value for the sample mean.

The probability-value approach is a third, equivalent method of hypothesis testing. The normal-curve table shows the probability of different z values. In our example, we had a z value of 2.05. Table 10.1 indicates that there is only a .02 chance of obtaining a z value above 2.05 by chance. There would also be .02 chance of obtaining a z value

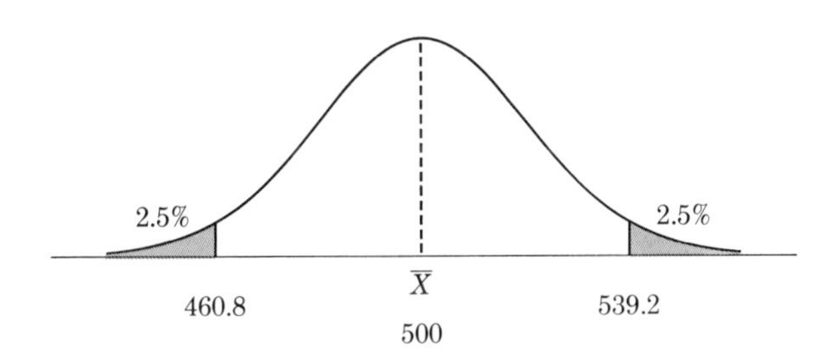

Figure 10.4
Critical Region at the .05 Level

below -2.05 by chance. So the overall probability of a sample mean at least 2.05 standard error units from the population mean would be $.02 + .02 = .04$. That would be the probability value, or *p* value.

If we decided to use the .05 significance level, we would reject the null hypothesis, since the *p* value is less than .05. (Another interpretation of the *p* value is that it is the probability that the null hypothesis is true. If the *p* value is less than .05, we would say that the result is significant at the .05 level.) This is entirely consistent with the results of the critical-ratio and critical-region procedures. However, the *p* value permits other researchers who believe the .01 level is more appropriate to decide not to reject the null hypothesis. The probability-value procedure relieves some of the arbitrary character of significance levels, and so we consider it the preferred method of hypothesis testing.

Conclusions

Note that if our data did not contradict the null hypothesis we would simply accept the null hypothesis. We would never prove the null hypothesis, because we can never prove that the population-mean GRE is 500 (or anything else) on the basis of a sample. At most, we can say that the sample is not inconsistent with such a possibility. Science proceeds by a series of disproofs rather than proofs. (This is because of the rules of logic referred to earlier.) This is part of the reason that Type II error is deemphasized. After all, if the null hypothesis really is wrong, a later study is likely to disprove it conclusively.

Many researchers are dissatisfied with the asymmetry between the two types of error. Others desire more information about the population mean than just whether or not an arbitrary null hypothesis is rejected. As a result, hypothesis testing is not very popular in some social-science fields. An alternative procedure for statistical inferences is establishing confidence intervals for the unknown population mean.

CONFIDENCE INTERVALS

If we are willing to accept a 5 percent chance of making an error, we can construct a confidence interval in which we can be 95 percent sure

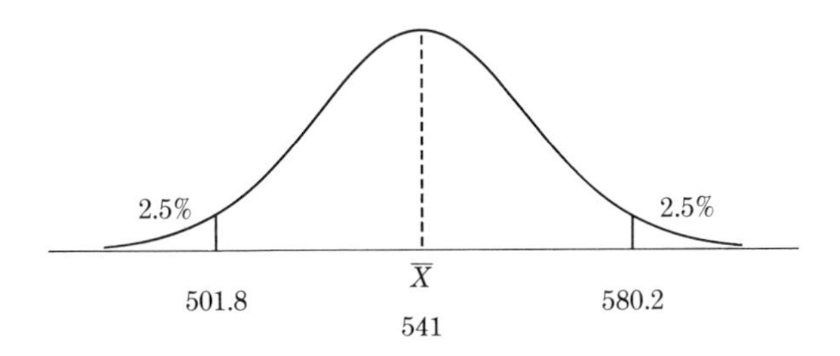

Figure 10.5
Ninety-Five Percent Confidence Interval for the Mean

that the true population mean falls. At least, if we took a large number of random samples and constructed confidence intervals for all of them, 95 percent of them would contain the population mean. The 95 percent confidence interval extends 1.96 standard error units either side of the sample mean.

In our example, the sample mean for your university was 541 and the standard error was 20. Therefore, the confidence interval would range from 501.8, or 541 − (1.96 × 20), to 580.2, 541 + (1.96 × 20). On the basis of our sample of only 26 cases, we would be able to conclude at a high (95 percent) level of confidence that the population-mean GRE for your university would be between 501.8 and 580.2 (see Figure 10.5).[2]

The calculation of the confidence interval may seem similar to the calculation of the critical region, and it is. Both involve a radius of 1.96 times the standard error. The critical region was centered around the null-hypothesis value, 500 in our example, and indicated which sample means (those below 460.8 and those above 539.2) would lead to rejection of the null hypothesis. We do not require a null hypothesis for the confidence interval. It is instead centered around the sample mean, 541 in our example. It indicates the region (501.8 to 580.2) in which the population mean is very likely (95 percent chance) to fall. Note that our earlier null-hypothesis value of 500 does not fall within the confidence interval. The two procedures always lead to identical conclusions with the type of test described so far. That is not the case in making "directional tests," as described in the next section.

[2]Because the national mean of 500 does not fall into this 95 percent confidence interval, we see once again that the mean for your school is significantly greater than the national mean.

DIRECTIONAL TEST

The hypothesis-testing procedure considered above has been based on the possibility of the sample mean being either above or below the hypothesized mean. In some situations, however, the researcher is not interested in one of those possibilities. Say, for example, that the GRE is known to have a mean of 500 for the population of college seniors. However, our sample involves students from elite universities, and we are interested in whether their abilities are significantly greater than those of the population. In any case, we have no concern that the mean for the population of students from which our sample was selected was less than 500. We want to test the null hypothesis that the mean for the population of elite universities was 500 (or less) against the research hypothesis that the mean is greater than 500.

Our previous significance testing involved nondirectional tests ("two-tailed tests") in which we were interested in whether our sample deviated significantly in either direction from the population mean. We looked at the probability of a sample mean at least 1.96 standard errors less than the population mean as well as the probability of a sample mean 1.96 or more standard error units above the population mean, and the probability of these two extreme conditions together was .05. In directional (or "one-tailed") tests, only one of these possibilities is of interest. There would be a .025 chance of obtaining a sample mean 1.96 standard error units above the population mean. So if we want to operate at the .05 level of significance, we would have to modify the 1.96 value. The normal-curve table indicates that there would be a .05 chance of obtaining a z value greater than 1.645. Therefore, we substitute $+1.645$ for $+1.96$ and -1.96 in directional significance tests at the .05 level.

We could use any of our earlier hypothesis-testing procedures for a directional test. First, we can calculate a critical ratio and check to see if it is greater than 1.645. Second, we can establish a critical region greater than 1.645 standard error units above the null hypothesis mean. Third, we can use the normal-curve table to find the probability of a z value more extreme than our obtained z value and then check to see if it is less than .05. (If we chose instead to construct a 95 percent confidence interval, we would still construct it 1.96 standard error units around the sample mean. The confidence interval would not be identical to directional hypothesis testing.)

Since the sample mean of 541 in our example above was significant at the .05 level with a nondirectional test, it would also be significant with a directional test. (If the z value or critical ratio was greater than 1.96, then it certainly is greater than 1.645.) Now assume our sample-mean GRE had been 535. We indicated in our discussion of the critical region that we would not have rejected the null hypothesis with a sample-mean GRE of 535 for a nondirectional test. What about a directional test? For 535, the z value would be 1.75. That is greater than 1.645; therefore, we would reject the null hypothesis that the population mean is 500 (or less). With a sample of 26, a sample-mean GRE of 535 would be enough to show that there is a 95 percent chance that the population mean is greater than 500. The critical region (Figure 10.6) for this test would involve rejecting the null hypothesis for sample means above 532.9, or 500 + (1.645 × 20). If the null hypothesis were true, the probability of our sample-mean GRE of 535 would be .04 (the value in Table 10.1 for a z of 1.75); hence we would reject the null hypothesis at the .05 level.

This example shows that means that are not significant in nondirectional tests can be significant in directional tests. Nondirectional tests are clearly more conservative. It would be fallacious to use the more lenient directional tests when you had no idea of which direction to test until you saw whether the sample mean was above or below the hypothesized mean. However, in some substantive situations one direction may make no sense, or we are certain of the actual direction, in which case directional tests should be used.

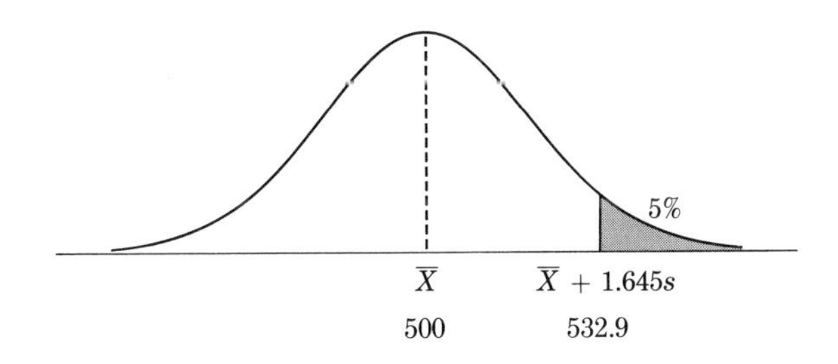

Figure 10.6
Critical Region for Nondirectional Test

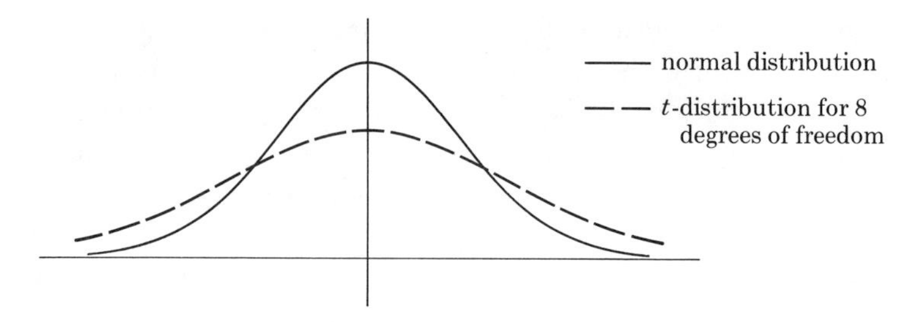

Figure 10.7
A Comparison of the *t*- and Normal Distribution

THE *t*-DISTRIBUTION

The previous analysis has made one very unusual assumption: that the population standard deviation was known. We rarely are in a situation to know the population standard deviation. At best, we can estimate it on the basis of our sample. However, the sampling distribution of means is not exactly normal when dealing with an estimated standard deviation. Instead it has what is known as the "*t*-distribution." That distribution has a shape (see Figure 10.7) that looks very much like the normal distribution, except that extreme values are slightly more likely. To complicate matters further, there is not one *t*-distribution but several, depending on the sample size. We define the "degrees of freedom" for a *t*-distribution as the sample size minus one. In significance testing and for establishing confidence intervals, we would actually use the *t*-distribution for the appropriate degrees of freedom instead of the normal distribution. It happens, though, that with a large number of cases (degrees of freedom) the *t*-distribution is virtually identical to the normal distribution. Therefore, the normal distribution can still be used for large samples, say of size 120 or greater. We need switch to the *t*-distribution only for small samples.

We now return to our above example of the nondirectional null hypothesis of a mean GRE of 500. Say that a sample of size 26 leads to a mean of 541 with an estimated standard error of 20. In that case, the appropriate test would involve the *t*-distribution based on $N - 1 = 25$ degrees of freedom. Table 10.3 shows that a value of 2.06 would be required for significance in this instance rather than the 1.96 used for the normal distribution. Our t value would be $2.05 = (541 - 500)/20$,

Table 10.3 Critical Ratios for the t-Distribution. (Entries show the t ratio required for significance.)

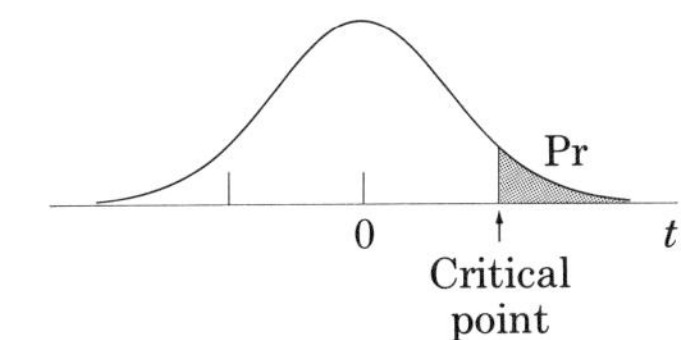

Pr d.f.	.05 Directional	.05 Nondirectional	.01 Directional	.01 Nondirectional
1	6.314	12.706	31.821	63.657
2	2.920	4.303	6.965	9.925
3	2.353	3.182	4.541	5.841
4	2.132	2.776	3.747	4.604
5	2.015	2.571	3.365	4.032
6	1.943	2.447	3.143	3.707
7	1.895	2.365	2.998	3.499
8	1.860	2.306	2.896	3.355
9	1.833	2.262	2.821	3.250
10	1.812	2.228	2.764	3.169
11	1.796	2.201	2.718	3.106
12	1.782	2.179	2.681	3.055
13	1.771	2.160	2.650	3.012
14	1.761	2.145	2.624	2.977
15	1.753	2.131	2.602	2.947
16	1.746	2.120	2.583	2.921
17	1.740	2.110	2.567	2.898
18	1.734	2.101	2.552	2.878
19	1.729	2.093	2.539	2.861
20	1.725	2.086	2.528	2.845
21	1.721	2.080	2.518	2.831
22	1.717	2.074	2.508	2.819
23	1.714	2.069	2.500	2.807
24	1.711	2.064	2.492	2.797
25	1.708	2.060	2.485	2.787
26	1.706	2.056	2.479	2.779
27	1.703	2.052	2.473	2.771
28	1.701	2.048	2.467	2.763
29	1.699	2.045	2.462	2.756
30	1.697	2.042	2.457	2.750
40	1.684	2.021	2.423	2.704
60	1.671	2.000	2.390	2.660
120	1.658	1.980	2.358	2.617
∞	1.645	1.960	2.326	2.576

SOURCE: Based on Thomas H. Wonnacott and Ronald J. Wonnacott, *Introductory Statistics,* 2nd ed. (New York: Wiley, Copyright © 1972), p. 481. Reprinted by permission of John Wiley & Sons, Inc. and Biometrika trustees.

which is less than the 2.06 required for significance at the .05 level. Therefore, we would not reject the null hypothesis. On the basis of the data, we cannot reject the possibility that the mean GRE is 500. Table 10.3 also shows that a value of 1.71 would be required for a di-reactional test for 25 degrees of freedom. Because our t value (2.05) in this instance is greater than the value required at the .05 level (1.71), we would reject the null hypothesis for the directional test. Hence, we can conclude that the mean GRE is greater than 500.

SUMMARY

If all of this seems complicated, the procedure is actually fairly straightforward in practice. First, consider whether the sample size is large (120 or greater). If it is, use the normal distribution. If it is small, check to see if the population standard deviation is somehow known. If it is, use the normal distribution. If the sample size is small and the population standard deviation is not known, use the t-distribution. Next decide whether to use a directional or nondirectional test. Nondirectional tests are preferred, because they are more conservative, but there are situations where substantively one direction makes no sense, so the test should be directional. Then determine the value required for significance with the distribution that you have chosen for that type of test.

As a final warning, if several significance tests are run, there would be a high likelihood that at least one would be significant by chance alone. The person who runs 100 significance tests and reports the 5 that turn out to be statistically significant at the .05 level is forgetting that 5 out of 100 tests will be significant by chance alone. Looking at all the possible tests to see which are significant can thus lead to nonsense.

We have described the significance procedures for means in great detail in this chapter. Similar procedures exist for other statistics, but the only other one that we shall present in any detail involves differences between means for different groups. That is the topic of Chapter 12.

Questions

1. According to a normal-curve table, what proportion of the area under the normal curve is above $z = .50$?
2. Say that a variable has a normal distribution with a mean of 75 and a standard deviation of 10.
   a. What is the probability of a value being above 95?
   b. What is the range within which 95 percent of the cases will fall?
3. Say that a variable has a normal distribution with a mean of 75 and a standard deviation of 10, and say that a sample of size 101 is taken.
   a. What is the standard error of the mean?
   b. Would a sample mean of 78.5 be significantly different from a hypothesized mean of 75 at the .01 level?
   c. What is the 99 percent confidence interval for a sample mean of 78.5?
4. Say that a variable has a normal distribution with a mean of 75 and a standard deviation of 10, and say that a sample of size 26 is taken.
   a. What is the standard error of the mean?
   b. Would a sample mean of 78.5 be significantly different from a hypothesized mean of 75 at the .05 level?
   c. What is the 95 percent confidence interval for a sample mean of 78.5?
5. Say that a variable has a *t*-distribution with a mean of 75 and a standard deviation of 10, and say that a sample of size 26 is taken.
   a. What is the standard error of the mean?
   b. Would a sample mean of 78.5 be significantly different from a hypothesized mean of 75 at the .05 level?
   c. What is the 95 percent confidence interval for a sample mean of 78.5?

11

Interval Statistics

This chapter will deal with the analysis of relationships among variables that are assumed to be measured at the interval level of measurement. The discussion of single-variable statistics in Chapter 5 already has presented the appropriate interval-level measures of central tendency and dispersion; the mean, the variance, and the standard deviation. In this chapter, we shall fill in the interval analogues of the techniques we have discussed in Chapters 7 through 9: measures of relationship, controls, and scaling.

THE ROLE OF INTERVAL STATISTICS

Before presenting specific techniques, we should call attention to the increasingly major role of interval statistics in social-science research. The techniques for dealing with interval-level data are generally much more powerful than those available for ordinal and nominal data. They permit stronger statements about the effects of one variable on another, and they greatly facilitate controlling for additional variables. Consequently, many researchers prefer interval-level analysis to the extent to which they have a choice. As a result, there has been a greater emphasis on collecting interval data as well as a greater interest in applying interval statistics to noninterval data.

Interval Data

Much of the data in particular subfields of the social sciences are interval-level data. For example, those studying comparative politics and international relations often work with national-attribute variables such as gross national product (GNP), population, defense expenditures, trade volume, and so on. All of these variables are measured at the interval level. Actually, these are just a few of the wide range of phenomena that can be measured at the interval level. Common variables that are routinely measured at the interval level include age, income, and education (years of schooling). Also, some subject domains for surveys naturally entail the interval level, as in surveys of family expenditure patterns (budget allocations for rent, food, and so on).

Additionally, researchers have learned to collect numeric data whenever they have the choice. For example, rather than just classify countries as rich or poor, they will collect the actual GNP's of the countries. Moreover, researchers have developed new procedures to obtain interval measurement. In particular, survey researchers have made attempts to move attitude measurement up to the interval level, such as by measuring attitudes toward presidential candidates with the feeling thermometer. Several of the scaling techniques discussed in the preceding chapter also seek to provide interval measurement.

Noninterval Data

A large segment of social-science research is concerned with what are clearly interval-level data, and, of course, the techniques discussed in this chapter are most appropriate for such data. However, many social scientists whose data are not interval-level nevertheless find interval statistics useful. One of the most common applications of interval statistics to what are usually ordinal or nominal data is in the field of survey research. The procedure for interval-level analysis of nominal data is straightforward and will be explained later in this chapter (p. 160).

The critical question is how great the risk of faulty conclusions is in analyzing ordinal variables with interval statistics. Some authorities now claim that too much has been made of the distinction between interval and ordinal measurement and that there is not much risk in this situation.[1] This point is controversial[2] (and there is even some chance of finding a negative correlation when the true correlation is positive, or vice versa).

Although this topic is still the subject of lively dispute, there is no doubt that analysts have been increasingly using interval-level analysis on ordinal data. As for ourselves, we remain very uneasy about this development, although it appears that the practice will generally not lead to faulty conclusions. In any case, interval analysis of interval and noninterval data is becoming so widespread that we feel that a comprehensive introduction to interval statistics is required.

CORRELATION AND REGRESSION

As we have indicated, interval-level techniques are much more powerful than ordinal-level and nominal-level techniques. They permit us not only to measure how strongly related a pair of variables are but also to make a statement describing the effect of one variable on the other; by means of such techniques, we can numerically measure the effect that a change in the independent variable has on the dependent variable. The associated changes are not always steady; consequently, we often focus on the *average* change in the dependent variable that occurs with a *constant* change in the independent variable. A "linear model" is particularly appropriate for this purpose and will be described in this chapter.

Linear Regression

Figure 11.1 shows a hypothetical relationship between education and income. Each point in this graph represents one survey respondent. Note the exact linear relationship: all of the points are on a single straight line. And it is a positive relationship, in the sense that those with higher amounts of education make greater incomes.

---

[1]Sanford Labovitz, "The Assignment of Numbers to Rank Order Categories," *American Sociological Review, 35* (1970), pp. 515–24.

[2]See, for example, David M. Grether, "Correlations with Ordinal Data," *Journal of Econometrics, 2* (1974), pp. 241–6.

How can the relationship between education and income in Figure 11.1 be summarized? First, we can determine what effect a change in education has on income. According to the graph, an increase of a year in education corresponds to an $800 increase in income. (For example, those with thirteen years of education make $800 more than those with twelve years of schooling.) In statistical parlance, the $800 figure is known as the *regression coefficient*—the change in the dependent variable induced by a unit change in the independent variable. It can also be seen to be the slope (that is, the tangent) of the line—the increase in height corresponding to a one unit move to the right in the figure.

Additionally, we can specify the equation of the straight line. Those with zero years of education have an annual income of $4,000, according to the figure. Therefore, the equation of the line is:

$$\text{Income} = \$4{,}000 + \$800\,(\text{Years of education})$$

(Thus, the income of those with twelve years of schooling is $4,000 + 12($800) = $4,000 + $9,600 = $13,600.) Statisticians would call the $4,000 figure the *intercept* of the line—the point at which the value of the dependent variable (income) corresponds to a value of zero for the independent variable (education).

Finally, we can summarize how well a person's income can be predicted from his or her education. In Figure 11.1, perfect prediction is possible using a linear prediction scheme. Statisticians use a *correlation coefficient* to summarize the accuracy of prediction. The correlation is perfect (equal to 1.0) in Figure 11.1 because the points are all along the straight line. It would be zero if education did not help predict income, and it would be negative if greater education actually led to lower income. The statistic that is used to measure linear correlation is known as "Pearson's r." Its square (r^2) indicates the proportion of the variance of the dependent variable that is explained by linear prediction from the independent variable.

Perfect relationships such as that shown in Figure 11.1, are never found in real data. Figure 11.2 shows a more realistic example. As education increases, people generally have a higher income. Even though the relationship is not perfectly linear, we can still seek the best-fitting linear prediction rule; such a line has been drawn in Figure 11.2. The figure shows that an increase of a year in education leads to an average $800 increase in income, so the regression coefficient is the same as in Figure 11.1. The intercept is also $4,000, so the regression equation is

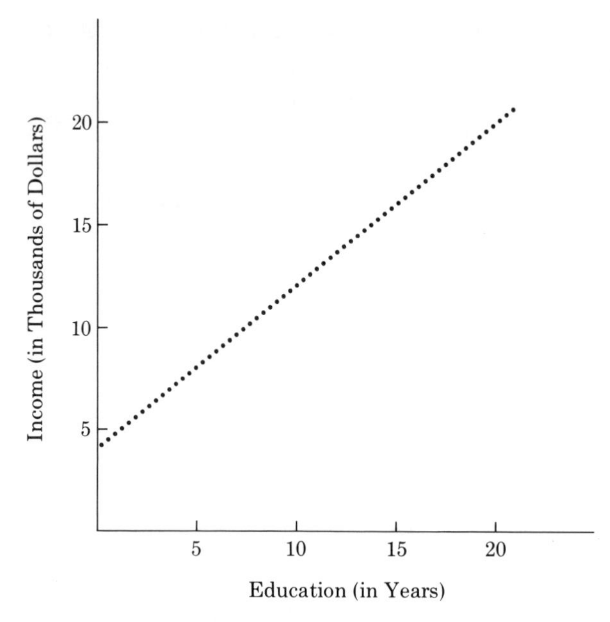

Figure 11.1
Hypothetical Linear Relationship
Between Education and Income

identical to that for Figure 11.1. However, the correlation between education and income is lower, since the relationship is no longer perfectly linear. There is still a positive relationship, but it is much weaker than in Figure 11.1.

The Procedure for Regression

How is the best-fitting linear prediction rule found? The rule will estimate the income value for a person with a given educational level. (For example, a person with twelve years of education is predicted to have an income of $13,600.) Yet, according to Figure 11.2, the predictions for many people will not be perfectly accurate. There is a deviation between the actual values and the predicted values. We want to find the prediction rule that *minimizes* these deviations, so that the predictions are as close to the actual values as possible. Mathematically, one effective means for minimization is to minimize the sum of the squared deviations. This is known as the "least-squares" criterion for the best-fitting linear relationship.

Notation. Before we can state this logic more formally, we must review the system of notation introduced in Chapter 5. Let us denote the dependent variable by Y and the score of person i on that variable by y_i (read "y-sub-i"). The mean value of Y (denoted as $\overline{Y}$, "Y-bar") is obtained by summing the set of Y values and dividing by the number of individuals, N. That is,

$$\overline{Y} = \frac{y_1 + y_2 + \cdots + y_N}{N} = \frac{\sum_{i=1}^{N} y_i}{N}$$

where $\sum_{i=1}^{N} y_i$ is the sum of the N observations on Y (literally it is read: "the sum of the y sub i's, where i goes from 1 to N").

All summations in this chapter are with respect to i, with i ranging from 1 to N (summing over the individuals), so we shall abbreviate this notation further to just $\sum y_i$ (or $\sum Y$). In this system of notation, the mean is

$$\overline{Y} = \frac{\sum y_i}{N}$$

and the variance of Y is

$$s_Y^2 = \frac{\sum (y_i - \overline{Y})^2}{N}$$

The Least-Squares Criterion. We shall now express the regression logic formally. Say that the dependent variable is labelled Y and that we wanted to predict person i's score on that variable (y_i). We can seek to predict that score on the basis of person i's score on some other variable (X) that is related to Y. Let y_i' be the value we predict for y_i on the basis of the person's X score (x_i). If we do not manage to predict person i's Y value exactly, there will be some error: ($y_i - y_i'$). We want this error to be as small as possible over the set of individuals. The "least–squares" criterion minimizes the average squared error:

$$s_{Y'}^2 = \frac{\sum (y_i - y_i')^2}{N}$$

The equation for a straight line can be written as:

$$y_i' = a + bx_i$$

where a is the intercept (where the line crosses the y axis when X equals zero) and b is the regression coefficient (or slope of the line). Algebra or calculus can be used to obtain formulas for the a and b values that lead to a prediction rule which minimizes $s_{Y'}^2$. Computers are generally used to compute a and b, but we shall work through an example of this calculation.

There are many ways of calculating the regression coefficient (b), and the constant (a). One formula for b is:

$$b = \frac{N\sum XY - \sum X \sum Y}{N\sum X^2 - \left(\sum X\right)^2}$$

When a person's X value equals the X mean, the best prediction for Y turns out to be the Y mean. So,

$$Y' = a + b\overline{X} = \overline{Y}$$

As a result, the estimate for the intercept is:

$$a = \overline{Y} - b\overline{X}$$

These formulas are illustrated in Table 11.1.

These formulas are fairly complex, and few social scientists do these calculations without the help of a pocket calculator or, more often, a computer. Although it is important to understand where the answers come from and how to interpret them, calculators and computers can spare us the burden of the actual computation. However, it is still up to us humans to analyze our data and to understand their meaning.

The Degree of Linear Relationship. How well does the linear regression equation fit the data? The residual variance, $s_{Y'}^2$, gives one indication, but another summary measure is of greater interest. We can determine the degree to which we can predict the Y values better knowing X than without knowing X.

If we had no knowledge of the person's score on possible predictor variables, our best guess of person i's Y score would be the Y mean, $\overline{Y}$. The prediction error for person i would be $(y_i - \overline{Y})$, and the average squared error over the set of individuals would be:

$$s_Y^2 = \frac{\sum (y_i - \overline{Y})^2}{N}$$

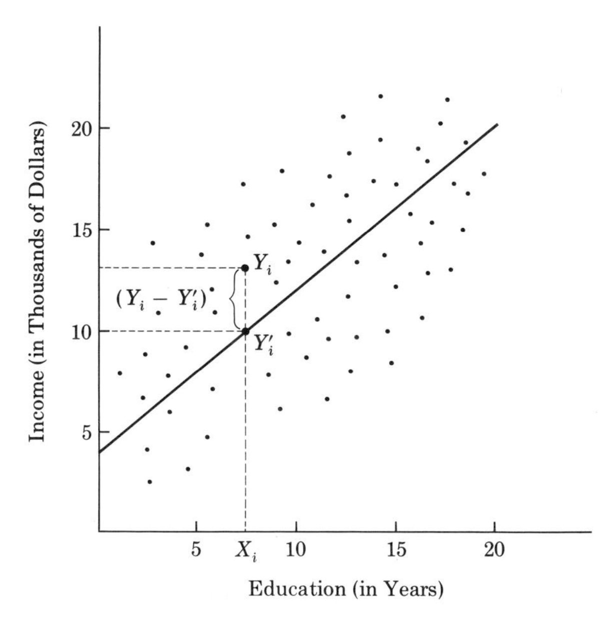

Figure 11.2
Hypothetical Imperfect Relationship
Between Education and Income

which is simply the variance of Y. Had we made any guess other than the Y mean, the average squared error would have been larger, so the mean is our "best estimate" of Y without knowledge of X. If s_Y^2 is the average squared error in predicting Y without knowledge of X and $s_{Y'}^2$ is the average squared error in prediction with knowledge of X, then the proportional reduction in error is:

$$r^2 = \frac{(s_Y^2 - s_{Y'}^2)}{s_Y^2}$$

where r turns out to be the correlation between X and Y.

Thus the square of the correlation coefficient can be interpreted as the proportion of the variance of Y that can be accounted for by linear prediction from X. (The proportional reduction in error notion is similar to that introduced for lambda in Chapter 7, except that there the number of category prediction errors were being reduced, and here the average squared numeric prediction error is being reduced.) An r^2 of 1.00 would indicate perfect prediction, while an r^2 of zero would show the independent variable does not help account for the dependent variable in a linear fashion.

This also provides us with an interpretation of Pearson r values. For example, if the correlation is above .7, more than half of the variance of the dependent variable can be accounted for by the independent variable. If the correlation is below .3, less than 10 percent of the variance of the dependent variable can be explained by the independent variable.

The correlation coefficient r can be calculated as:

$$r = \frac{N\sum XY - \left(\sum X\right)\left(\sum Y\right)}{\sqrt{\left[N\sum X^2 - \left(\sum X\right)^2\right]\left[N\sum Y^2 - \left(\sum Y\right)^2\right]}}.$$

This formula is also illustrated in Table 11.1.

Regression Residuals. Recall that the regression was designed to minimize the sum of squared deviations from the regression line. Table 11.2 continues the example of Table 11.1 by showing the Y values predicted from the regression equation of Table 11.1 and the deviations between the actual and predicted Y values. These deviations, the $(Y - Y')$ column, are often termed "residuals." Table 11.2 gives the $s^2_{Y'}$ value for these data and shows how the r^2 interpretation in terms of proportional reduction in error gives the same r^2 as found in Table 11.1.

These residuals are of interest because they indicate what part of the dependent variable has not been explained by linear prediction from the independent variable. If the dependent variable is to be further understood, we must find which other possible independent variables are correlated with these residuals. For example, if we found that persons 3, 6, 9, and 12 in Table 11.2 were older than the others and that persons 1, 4, 7, and 10 were younger than the others, then we would conclude that age is correlated with the residuals and therefore should be included as an explanatory variable. We continue with this particular example in the next section, but it is important to emphasize the role of "residual analysis" in deciding whether additional explanatory variables should be included in a model.

Summary. In summary, we usually measure the association of interval variables with respect to a linear-prediction model. The regression coefficient, b, is the rate of change of the dependent variable (Y) with respect to the independent variable (X); the correlation coefficient, r, measures how well the data fit the line described by a and b. The square of the correlation coefficient, r^2, can be interpreted as the proportion of variance of Y which can be accounted for by X.

Table 11.1 Computation of Correlation and Regression Statistics

Person	Age (X)	X^2	Income (Y)	Y^2	XY
1	6	36	$ 3,800	14,400,000	22,800
2	6	36	9,000	81,000,000	54,000
3	6	36	13,600	184,960,000	81,600
4	12	144	10,300	106,090,000	123,600
5	12	144	13,900	193,210,000	166,800
6	12	144	16,600	275,560,000	199,200
7	16	256	13,800	190,440,000	220,800
8	16	256	16,800	282,240,000	268,800
9	16	256	19,800	392,040,000	316,800
10	20	400	14,900	222,010,000	298,000
11	20	400	20,200	408,040,000	404,000
12	20	400	24,900	620,010,000	498,000
Sum	162	2,508	177,600	2,970,040,000	2,654,400

$$r = \frac{12(2{,}654{,}400) - (162)(177{,}600)}{\sqrt{12(2{,}508) - 162^2}\ \sqrt{12(2{,}970{,}040{,}000) - 177{,}600^2}} = \frac{3{,}081{,}600}{3{,}973{,}446} = .776$$

$$r^2 = .776^2 = .601$$

$$b = \frac{12(2{,}654{,}400) - (162)(177{,}600)}{12(2{,}508) - 162^2} = \frac{3{,}081{,}600}{3{,}852} = 800$$

$$\overline{X} = \frac{162}{12} = 13.5$$

$$\overline{Y} = \frac{\$177{,}600}{12} = \$14{,}800$$

$$a = \$14{,}800 - \$800(13.5) = \$4{,}000.$$

$$Y' = \$4000 + \$800X$$

Interpretation of Regression

Some Cautions. It should be emphasized that the correlation and regression coefficients are based exclusively on a linear model. Sometimes there will be a strong nonlinear relationship between a pair of variables. Imagine, for example, that people with high levels of education had difficulty in getting jobs because there were more such people looking for jobs than there were jobs suitable for them. The relationship might then look like the one shown in Figure 11.3. There is a strong *predictive* relationship between education and income, but it is not linear. The Pearson's r value here would be very small because all it measures is linear correlation. The possibility of

Table 11.2 Calculation of Regression Residuals

Person	Age (X)	Income (Y)	Y'	Residual $Y - Y'$	$(Y - Y')^2$
1	6	\$ 3,800	\$ 8,800	−\$5,000	25,000,000
2	6	9,000	8,800	200	40,000
3	6	13,600	8,800	4,800	23,040,000
4	12	10,300	13,600	−3,300	10,890,000
5	12	13,900	13,600	300	90,000
6	12	16,600	13,600	3,000	9,000,000
7	16	13,800	16,800	−3,000	9,000,000
8	16	16,800	16,800	0	0
9	16	19,800	16,800	3,000	9,000,000
10	20	14,900	20,000	−5,100	26,010,000
11	20	20,200	20,000	200	40,000
12	20	24,900	20,000	4,900	24,010,000
Sum	162	177,600	177,600	0	136,120,000

$$Y \text{ Variance} = s_Y^2 = \frac{12(2{,}970{,}040{,}000) - 177{,}600^2}{12^2} = 2{,}846.33$$

$$\text{Residual Variance} = s_{Y'}^2 = \frac{136{,}120{,}000}{12} = 1{,}134.33$$

$$r^2 = \frac{2{,}846.33 - 1{,}134.33}{2{,}846.33} = .601$$

"curvilinear" relationships makes it important to look at the graph of the relationship between the variables rather than relying exclusively on the correlation coefficient.

Graphing the data can also help detect some undesirable conditions. For example, there might be a strong apparent correlation between the variables only because of one or two measurements. In Figure 11.4, there is no correlation for the main cluster of people. However, two people have an unusually high education and an unusually high income, and this is enough to make the total correlation in Figure 11.4 large. If the computer is directed to graph the variables (such "scatter plot" or "scatter diagram" programs are contained in most major social-science programs), a glance at the graph would be enough to indicate whether an apparently high correlation is caused by a few "outliers" as in Figure 11.4.

Significance Tests. In previous chapters we indicated that sampling error should be taken into account when evaluating statistical results, and the same is true for correlation and regression

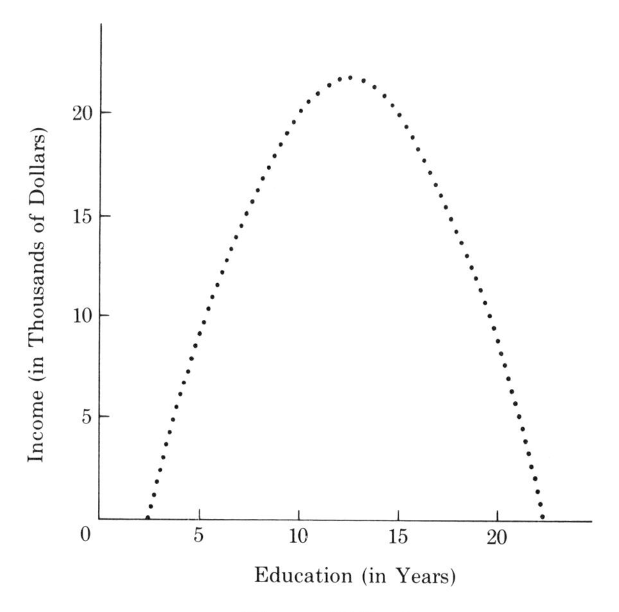

Figure 11.3
Hypothetical Curvilinear Relationship
Between Education and Income

coefficients. Yet, as before, we must bear in mind that the standard formulas are based on simple random sampling, which is not the most common sampling procedure for surveys. For those times when it is appropriate, we shall now outline the approach for taking sampling error into account for simple random samples.

The important question is whether the apparent correlation between the variables might be totally explained by sampling error—which is equivalent to asking whether the correlation coefficient differs significantly from zero. A correlation of zero would mean that the independent variable does not help predict the dependent variable, so this amounts to a test of whether the independent variable has a statistically significant effect on the dependent variable. If we have a simple random sample from a population and if we assume that the variables have normal distributions (the bell-shaped form of Figure 5.1), then Table 11.3 can be used to test the significance of the correlation coefficient, as a function of the number of respondents on which the correlation is based.

Table 11.3 Pearson r Values Required for Significance ($p = .05$)

Number of Cases	Nondirectional Test	Directional Test
10	.63	.55
15	.51	.44
20	.44	.38
25	.40	.34
32	.35	.30
42	.30	.26
52	.27	.23
62	.25	.21
72	.23	.195
82	.22	.18
92	.205	.17
102	.195	.16
152	.16	.13
202	.14	.12
302	.11	.095
502	.09	.07
1,002	.06	.05
2,002	.04	.04

If we wished to test whether a correlation were significantly *different from zero,* the "nondirectional test" column of Table 11.3 would show the minimum correlation value required for significance (taking no more than a 5 percent chance of declaring a correlation significant when the population correlation is actually zero). For example, a correlation of .20 would be significant for a sample of size 102, though a correlation of .19 would not be. In some situations, we might be sure that a correlation is positive. For example, even though we might not be sure whether education has a significant impact on income, we certainly expect that the effect of education on income would be positive. In that instance, the "directional test" column would be used to determine if the correlation were significantly *greater than* zero. This is a less rigorous test than the nondirectional (often called a "two-tailed") test, as can be seen by the fact that the .19 correlation is now significantly greater than zero for a sample of size 102. The directional (or "one-tailed") test should be used only if you were sure of the sign of the correlation coefficient *before* you looked at it.

Unfortunately, significance tests are not very useful for large samples, such as those used in most surveys. Recall that the square of the correlation coefficient indicates how much of the variance of the dependent variable has been explained statistically. A glance at Table

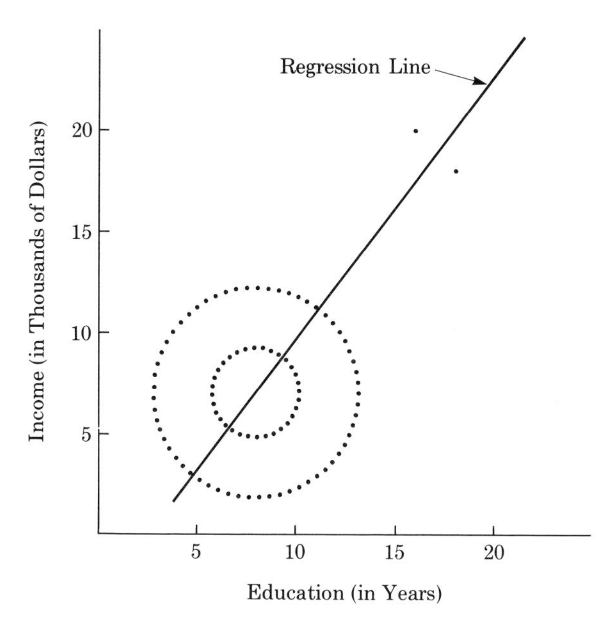

Figure 11.4
The Effect of Outliers on Regression

11.3 shows that with large samples a correlation may be statistically significant even though it does not account for much of the variance in the dependent variable. (Even if a correlation of .20 is "significant," only 4 percent of the variance is being explained!) As a result, the researcher should not rely on significance tests but should carefully consider whether the independent variable is really having a substantial impact on the dependent variable. For example, the correlation between the meaningless sequential number of each interview in the computer file and vote in the 1972 Center for Political Studies American National Election Study was .037, which will occur by chance with 1,582 cases only about 7 times out of 100; however, the correlation explains a trivial proportion of the variance on the vote (less than two-tenths of one percent of the variance) and has absolutely no substantive significance. By itself, statistical significance is not important for large surveys.

Finally, the nature of significance tests should be kept in mind even when evaluating a significant correlation. When a correlation is significant according to Table 11.3, there is no more than a 5 percent chance that the true population correlation is actually zero—in other words, if a large number of correlations are examined, some 5 percent of them will be significant by chance alone. If a researcher generates a hundred correlations and finds four, five, or six significant, even those significant correlations might still be due to chance.

An Example

As an example of the use of two-variable (simple) regression analysis with survey data, we will consider the 1972 CPS American National Election Study question regarding the proper role of women. The question asked the respondents to indicate their position on a seven-point scale, ranging from favoring an equal role for women (category 1) to favoring a traditional role for women (category 7). This variable could be considered ordinal, but we shall treat it as interval in order to illustrate the regression approach.

Since a "generation gap" might be expected on this question, age effects on attitudes toward the role of women were examined. The Pearson's r correlation between age (in years) and the women's role question is .07. This correlation is very small, but it is statistically significant since it is based on 2,685 respondents. The regression equation is: Attitude $= 2.938 + .010$ age. In other words, a year of age changes attitudes toward women by only .10. That effect seems very small, but it leads to an expected difference of .50 between 68-year-old people and 18-year-olds. Older people do have a more traditional image of woman's role in society than do younger people, but the difference is slight. Age explains less than one percent of the variance in attitudes toward women ($r^2 = .0056$). There is a generation gap, but it has very slight impact on these attitudes.

CONTROLS

Even if a correlation is both significant and sizable, correlation does not prove causation. Two variables might be correlated not because one causes the other but because both are caused by the same third variable. For example, why are education and income correlated? Perhaps because both are caused by the social class of the person's parents.[3]

[3]Assume that we have an interval-level measure of parent's social class.

Table 11.4 Spurious Relationship Between Education and Income

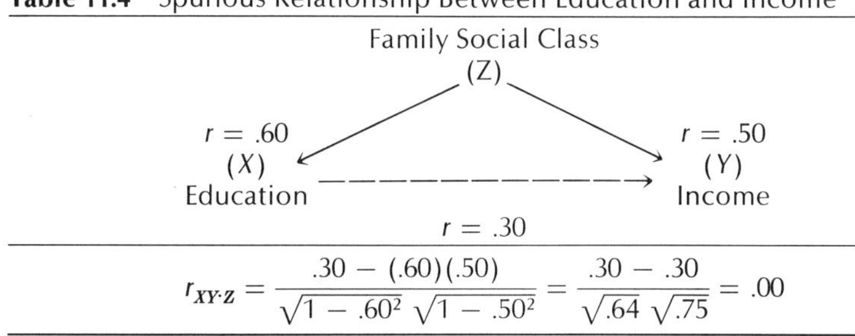

$$r_{XY \cdot Z} = \frac{.30 - (.60)(.50)}{\sqrt{1 - .60^2}\,\sqrt{1 - .50^2}} = \frac{.30 - .30}{\sqrt{.64}\,\sqrt{.75}} = .00$$

People who come from an upper-middle-class background may tend to have more education and greater incomes themselves; people who come from a working-class background may tend to have less education and a lower income. Table 11.4 illustrates how family social class could cause the apparent relationship between education and income. To determine whether family social class is the cause of the education–income relationship, we would like to look at the education-income relationship with the effects of family social class statistically removed.

Partial Correlation

How can the effects of a variable be removed from a relationship? The statistical-controls procedure described in Chapter 8 does so by dividing the sample into separate groups of respondents with the same measured value on this variable, so the relationship could be examined separately within each group. The regression logic explained in the previous section permits a different approach for interval data.

If we use regression to predict the person's education from the person's family social class, the residuals would indicate the part of the person's education that could not be predicted from family social class. Similarly, if we predict the person's income from his or her family social class, the residuals would indicate the part of the person's income that could not be predicted by family social class. Having removed the effects of the control variable (family social class) from both variables, we can recalculate the correlation between education and income. This correlation between residuals is known as a "partial correlation."

If education is labeled *X,* income *Y,* and family social class *Z,* then the partial correlation coefficient of interest is the correlation of *X* and

Y controlling for variable Z, denoted $r_{XY \cdot Z}$. It can be computed from the pairwise correlations:

$$r_{XY \cdot Z} = \frac{r_{XY} - r_{XZ}r_{YZ}}{\sqrt{1 - r_{XZ}^2}\sqrt{1 - r_{YZ}^2}}$$

In an analogous manner, we can obtain "higher-order partials" that remove the effects of more than one variable from a relationship, such as the relationship between education and income controlling for family social class and age of respondent.

Comparison with Statistical Controls

As we have shown, when the data are interval, controlling by physically separating the cases into groups as in Chapter 8 is not necessary. This is particularly useful when there are too few cases in one or more of the groups to produce a reliable measure of association.

This partialing technique works particularly well in determining whether a two-variable relationship is "spurious" (see Chapter 8). The education-income relationship would be spurious if it could be explained entirely in terms of the two variables having a common cause. In that case, the correlation between education and income would equal the product of the correlations of each of those variables with the third variable, as in Table 11.4. By the formula given above, the partial correlation between education and income controlling for the third variable would then be zero. Thus, if the original correlation between two variables is large but the partial between them controlling on a third variable is zero, the original relationship is shown to be spurious. If the third variable does not fully explain the two-variable relationship, then the partial correlation coefficient would indicate how related education and income are above and beyond their relationships with the third variable.

The partial-correlation coefficient is less useful when a third variable "specifies" a two-variable relationship (see Chapter 8). For example, say that the correlation between education and income was positive for older people but negative for younger people (because young people with college education are still in school or just starting their careers). The partial correlation coefficient cannot indicate that the correlations for separate control groups are very different. Instead, the partial correlation between education and income controlling on age might be near zero because it is a weighted average of the correlations for the different age groups.

MULTIPLE REGRESSION

The discussion of causal processes in Chapter 3 emphasized the possibility of several independent variables jointly causing a dependent variable. Perhaps, for example, income is a function of both a person's education and the person's age. An older person with a college education probably has greater job experience and seniority than a younger person with a college education, so the older person is likely to have a higher income. What we require is a prediction rule for income that takes into account the person's education and age. Table 11.5 illustrates this model.

The Procedure for Regression

We cannot merely calculate separate simple regressions to determine the effect of education on income and then the effect of age on income. (Education and age are likely to be correlated, so part of the apparent effect of education is really an age effect and vice versa.) Instead, we must determine the effect of education on income with age held constant and the effect of age on income with education held constant.

The above discussion of partial correlations should suggest that this could be done by regressing residuals. To determine the effect of education on income with age held constant, we could predict income from age, then predict education from age, and finally calculate a regression

Table 11.5 Income as a Function of Education and Age

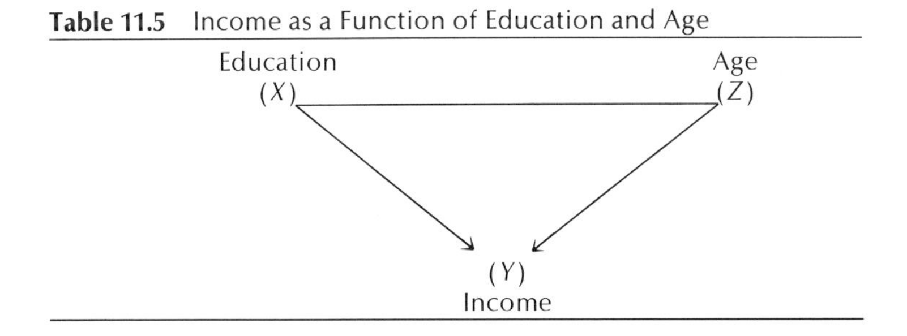

between the two sets of residuals. The resultant "partial regression coefficient" would give the effect of education on income controlling on age; similarly we could calculate the partial regression coefficient showing the effect of age on income controlling on education. The partial regression coefficients would show the rate of change of the dependent variable with respect to each of the independent variables, controlling on the other independent variables.

Rather than actually compute the residuals and regress them, "multiple linear regression analysis" is usually used to obtain the partial regression coefficients. Its computations are different from those described in the previous paragraph, but it gives identical results. Multiple-regression analysis seeks a linear-prediction rule for the dependent variable (which we label Y) from the independent variables (which we label X and Z), such that

$$Y' = a + b_{YX \cdot Z}X + b_{YZ \cdot X}Z$$

The a is an intercept term for the regression; the b's are partial slopes. (In each case, the first subscript denotes the dependent variable, the second the independent, and control variables are listed after the dot.)

As in the simple regression explained at the beginning of this chapter, the a and b's are chosen in such a manner that they predict the dependent variable Y with a minimum of squared error. That is, the "least squares" procedures seeks to minimize the average squared error:

$$s^2_{Y'} = \frac{\sum_{i=1}^{N} (y_i - y'_i)^2}{N}$$

We will not derive the a and b terms here nor show their formulas—these calculations are nearly always done on a computer.

As in two-variable regression, we can summarize the extent to which the dependent variable can be predicted from the independent variables with a correlation coefficient. The "multiple correlation coefficient" is denoted as R, and its square (R^2) gives the proportion of the variance in the dependent variable that can be explained by all of the independent variables acting together according to a linear rule. Or, in terms of proportional reduction in error,

$$R^2 = \frac{s^2_Y - s^2_{Y'}}{s^2_Y}$$

Interpretation of Multiple Regression

Comparing Predictors. An important question is: Which predictor is more important? This is often difficult to judge from the regression equation. In part, this is because the different predictors are sometimes measured in different units. In the present example, they are both measured in years, but that would obviously not be true if we were predicting income from education and parent's income. Additionally the different predictors have different amounts of variance. Even though education and age are both measured in years, the differences between people in education (from zero to twenty years) is much less than the differences between them in age (from twenty-five to sixty-five). A year of age may seem to make less difference than a year of education, but forty years of age could have a greater effect than twenty years of education.

What is required is a means of comparing the impact of variables measured in different units and with different variances. This is done by *standardized regression coefficients,* often known as *beta weights* (β). Each variable is given the same variance by subtracting the mean for the variable and dividing by its standard deviation. These standardized variables now have a mean of zero and a variance of one. Once the variables all have the same variance, new regression coefficients are calculated; these coefficients indicate the relative importance of the variables. In practice, the regression coefficient need not actually be recalculated but merely adjusted by the computer. The difference between unstandardized and standardized values depends on the variances of the variables:

$$\beta_{YX \cdot Z} = b_{YX \cdot Z} \frac{s_X}{s_Y}$$

Significance Tests. The significance of the regression coefficients is easy to test. Again this assumes simple random sampling and a normal distribution underlying the data. Computer programs for regression routinely print out the "standard error" of the regression coefficients. The usual test (taking 5 percent chance of error) is whether the coefficient is at least twice its standard error. (This is often expressed in terms of a "t statistic," which should generally be at least 2, or an "F statistic," which should generally be at least 4.) More exact probability statements can be made, but this rule of thumb is actually very close for regressions with at least sixty cases. If a coefficient is not significantly different from zero, then that variable can be safely dropped from the regression. The multiple-correlation coefficient also has a significance test. If it is not significantly above zero, then the regression exercise has not helped explain the dependent variable.

Some Cautions. The regression approach explained here actually can encounter a large variety of problems. Linear regression cannot detect the curvilinear form of the relationship between a pair of variables shown in Figure 11.3. It also does not detect "interaction effects" of combinations of independent variables (see pages 99–100 and 192), as when education has a greater impact on the income of younger people than of older people. It is possible to use more complex regression approaches, which can take even these factors into account, but doing so requires careful statement of one's theory in advance.

Another potential problem is that the predictors may be so highly correlated that their separate effects cannot be distinguished. To use our example, the most extreme case would be if age and education were perfectly correlated, so that older people always had more education than younger ones; the effects of age and education could not be separated. We never find perfect correlation, but if two predictors are correlated at more than .70, their regression coefficients become so unstable that we cannot rely on our estimates of them. Technically, this condition is referred to as "multicollinearity."

Dichotomous Variables. Special questions are raised by dichotomous variables. They can be used as predictors by scoring one category as "1" and the other by "0." This is known as a "dummy variable." It is also possible to handle nominal variables in a similar way. For example, if there are four categories to region (north, east, south, and west), then three dummies can be constructed: (1) north versus (0) rest; (1) east versus (0) rest; and (1) south versus (0) rest. (Why not west versus rest? Because the west is the only region with the score of 0 on all the other variables, so its effect is actually the base line.)

What about treating dichotomous variables as dependent? The problem is that although the regression approach will predict a large number of values, the dependent variable can only attain the values of 0 and 1. This will inevitably detract from the quality of the regression. The conventional statistical advice is to avoid regressions with dichotomous dependent variables. However, increasingly such regres-

Table 11.6 A Regression Analysis of Attitudes Toward the Role of Women

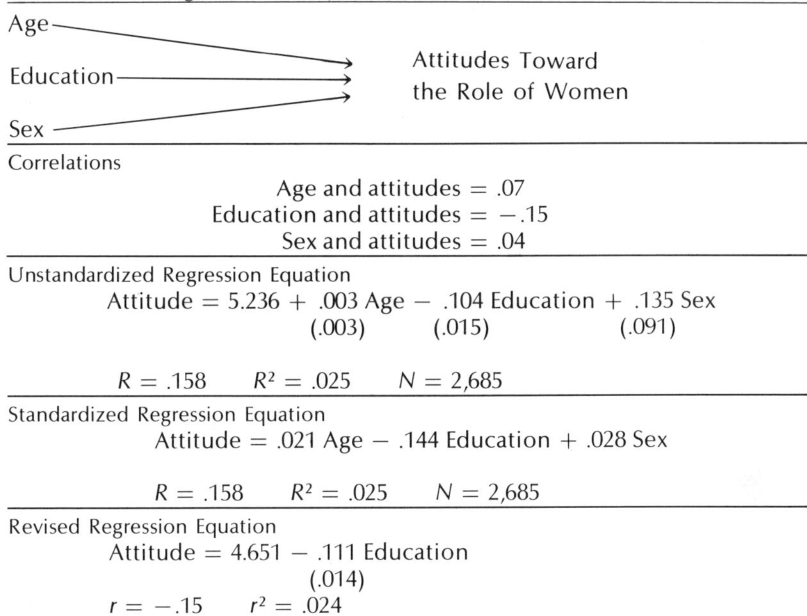

Age → Attitudes Toward the Role of Women
Education → Attitudes Toward the Role of Women
Sex → Attitudes Toward the Role of Women

Correlations

Age and attitudes = .07
Education and attitudes = −.15
Sex and attitudes = .04

Unstandardized Regression Equation

$$\text{Attitude} = 5.236 + \underset{(.003)}{.003}\ \text{Age} - \underset{(.015)}{.104}\ \text{Education} + \underset{(.091)}{.135}\ \text{Sex}$$

$R = .158 \qquad R^2 = .025 \qquad N = 2{,}685$

Standardized Regression Equation

$$\text{Attitude} = .021\ \text{Age} - .144\ \text{Education} + .028\ \text{Sex}$$

$R = .158 \qquad R^2 = .025 \qquad N = 2{,}685$

Revised Regression Equation

$$\text{Attitude} = 4.651 - \underset{(.014)}{.111}\ \text{Education}$$

$r = -.15 \qquad r^2 = .024$

NOTE: Figures in parenthesis are the standard errors of the corresponding regression coefficients.
SOURCE: Center for Political Studies, 1972 American National Election Study.

sions are being done. Alternative regression techniques for getting around this do exist, though they are beyond the scope of this introductory book.

An Example

As an example of multiple regression with survey data, Table 11.6 resumes the analysis of the 1972 election study question regarding attitudes toward the role of women. High scores on the dependent variable indicate support for a traditional women's role. The independent variables are age, years of education, and sex (coded as a dummy variable, 0 for men and 1 for women). According to the correlations, older people, people with less education, and women are more likely to support a traditional role for women than are others.

As indicated earlier in this chapter, the age effect is small but statistically significant. However, the age effect is eliminated when education is added to the prediction equation. Comparing the partial regression coefficients with their standard errors, only the education effect is statistically significant. Looking at the beta weights, the education effect is six to seven times more important than the sex and age effects. There is an "education gap," but the previously discovered generation gap disappears when education is brought into the analysis.

Since sex and age effects are so small, Table 11.6 also includes a recalculation of the regression using education alone as a predictor. According to the prediction equation, people without any formal education would be expected to have an average score of 4.65 (towards the traditional end of the scale) while those with twenty years of education would be predicted to have a score of 2.42 (towards the equal role end). Yet this analysis accounts for only about 2 percent of the variance in the dependent variable. Other variables must be brought into the analysis if we are to understand the determinants of attitudes toward the role of women.

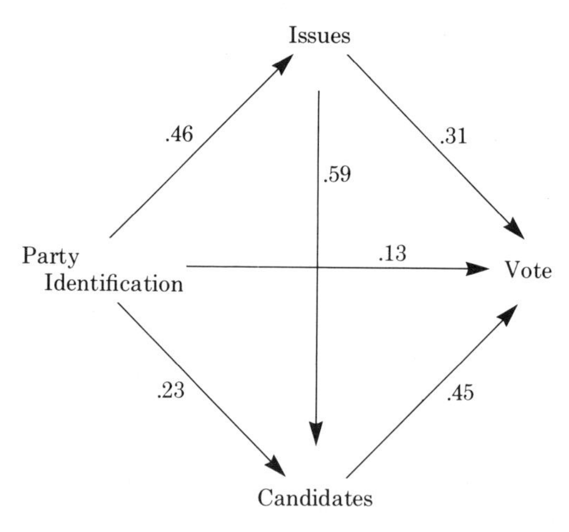

Total Effects
Party Identification = .49
Issues = .58
Candidates = .45

Figure 11.5
A Causal Model of Voting for 1972

(From Arthur H. Miller, Warren E. Miller, Alden S. Raine, and Thad A. Brown, "A Majority Party in Disarray," *American Political Science Review, 70,* 1976, forthcoming, fn. 29, Reprinted with permission of publisher.)

Causal Modeling

The last complication we shall bring into this introduction to multiple regression is the notion of the complex causal process, which we mentioned in Chapter 3. If the causal process has several elements, then a series of regression equations is required. There are simultaneous-equation methods to handle such cases, and *causal-modeling* and *path-analysis* procedures for determining the relative importance of the various causal paths.

An example of this approach is shown in Figure 11.5. This is a causal model of the vote, though a simpler one than the model with reciprocal causation between issue opinions and candidate evaluations in Figure 3.1. The "path coefficients" along the arrows are similar to partial regression coefficients. Candidate orientation has the greatest *direct* effect on vote for 1972 in this model (.45), though, since it affects the vote both directly and indirectly, issue position has the greatest *total* effect (.58).[4]

FACTOR ANALYSIS

A final procedure important for analysis of interval variables is *factor analysis*. In many respects it closely resembles the multidimensional-scaling technique described in the previous chapter. The basic purpose of factor analysis is to reduce or simplify the interrelationships among a set of variables. It is a complicated mathematical procedure that requires the use of a computer. Generally, one starts with a correlation matrix that contains the correlations of each variable with all of the other variables. Factor analysis can be thought of as a geometric representation of these correlations.

The Geometric Interpretation of Factor Analysis

One way of looking at the factor-analysis process is to imagine the dependent variables as lines of a specific length drawn from a com-

[4]For an explanation of the measures used here, see Arthur H. Miller, Warren E. Miller, Alden S. Raine, and Thad A. Brown, "A Majority Party in Disarray," *American Political Science Review, 70* (1976), forthcoming.

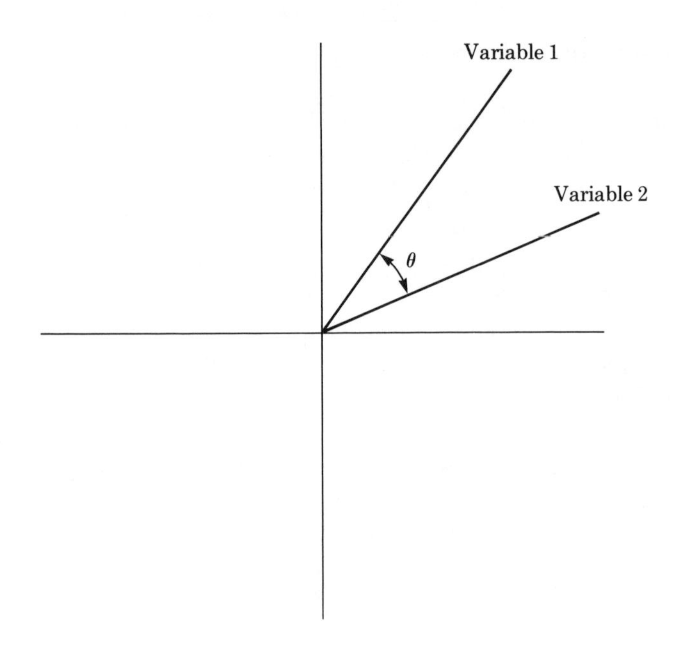

Figure 11.6
Angular Interpretation of a Correlation

mon point of origin in a multidimensional space, the dimensions, or factors, being the unknown underlying independent variables. (See Figure 11.6.) The lines are drawn one at a time so that the angle (θ, theta) that each variable makes with each other variable that is already in the space is related to its correlation with that variable. If two variables were perfectly correlated, the angle between them would be zero and they would appear as a single line. On the other hand, two variables that are not related to one another ($r = 0$) would be placed at a 90° angle. In supplying a spatial representation of the correlations, factor analysis tells us how many dimensions are necessary in order to account for the relationships in the correlation matrix. This number can be interpreted as a measure of how simple or complex those relationships are. As successive variables are added, it will become increasingly difficult to add variables to the space without increasing the number of dimensions. It is always possible mathematically to keep adding dimensions and to account for n variables in an n-dimensional space, but factor analysis is considered successful only if the number of dimensions is small enough to simplify the interpretation of the data.

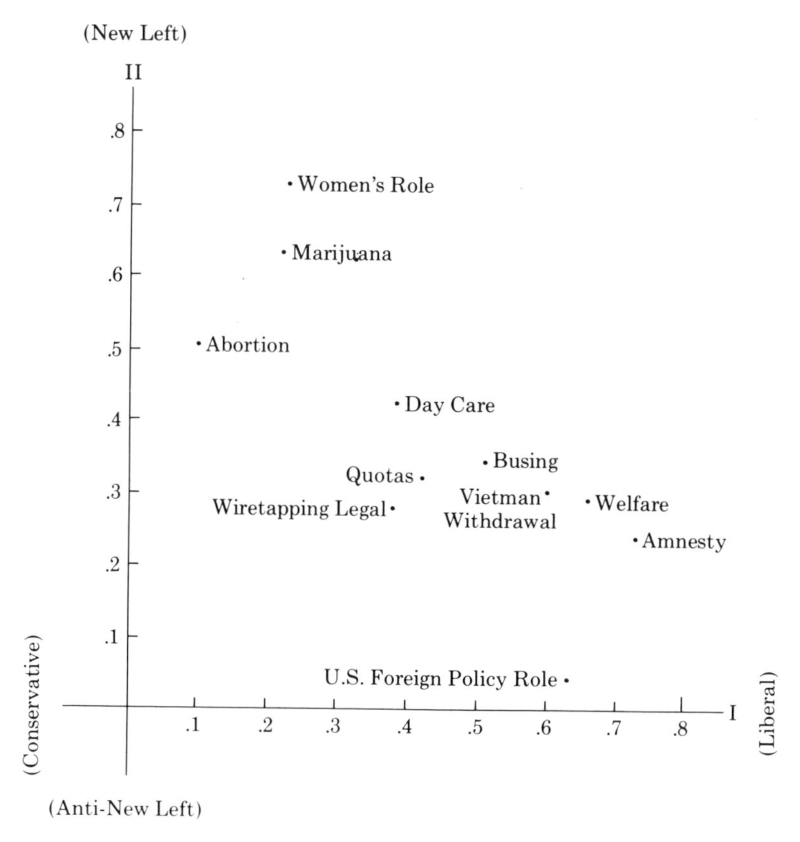

Figure 11.7
Factor Analysis of Ann Arbor Attitudes

An Example

Figure 11.7 gives an example of a factor-analysis solution. The data are from a 1972 survey of voters in Ann Arbor, Michigan. Factor analysis found that only two factors are required to account for the correlations among eleven issue questions. The figure shows how these issue questions can be explained by the two factors. One interpretation of the solution might be that the horizontal axis (Factor I) is a traditional left-right (liberal-conservative) continuum and that the vertical axis (Factor II) is a new-left/anti-new-left continuum. Note that issues like women's role, abortion, and marijuana are closely related to the new-left dimension and issues like the foreign policy role of the U.S.,

Table 11.7 Factor Loadings for Issue Questions in the 1972 Ann Arbor Study (Varimax Rotation)

Issue Questions	Factor I	Factor II	Communality
Day Care	.376	.421	.319
Traditional Women's Role	.233	.718	.570
Welfare	.655	.267	.500
Busing	.498	.328	.356
Amnesty	.728	.239	.587
U.S. Foreign Policy Role	.633	.046	.403
Vietnam Withdrawal	.578	.292	.419
Abortion	.117	.507	.271
Quotas for Minorities	.411	.326	.275
Legalize Marijuana	.220	.624	.438
Wiretapping Legal	.360	.262	.198
Factor Contributions	2.498	1.837	4.336
Percent of Total Variance	22.7%	16.7%	39.4%

welfare, and amnesty are more closely related to the traditional left-right continuum. Busing and quotas for minorities represent issues that are almost equally represented on both dimensions.

Table 11.7 shows the factor loadings (the correlation between the factors and the variables) for all of the issue questions and both factors. (These are also the coordinates of the variables for Figure 11.7.) Not only might it be necessary to add a third dimension to the space to account for the correlations, but it might be necessary to add a fourth, fifth, or even sixth dimension. Although we may not be able to visualize the entire space at once (without mind-expansion exercise or assistance), we can still look at plots of pairs of factors. There are no physical (length, width, height) interpretations of dimensions past the third, but the mathematical properties of these dimensions are identical to those of the familiar first three dimensions.

The column headed "communality" indicates the amount of the variance of the variable that is explained by the factor analysis. (This also shows the squared length of the line that represents the variable in the factor space, the squared distance of each point from the origin in Figure 11.7.) For example, 50 percent of the variance on the welfare issue is explained by the factor analysis, but only about 20 percent of the variance on the wiretapping issue is explained. The bottom row of the table shows that the first factor accounts for just 22.7 percent of the total variance of the eleven questions (2.498/11) and that the first two factors together account for only 39.4 percent of the total variance. Much of the variance remains unexplained by this analysis.

INTERVAL-ANALYSIS TECHNIQUES

Some researchers are addicts of factor analysis and will employ the procedure at every opportunity; other researchers seem to be addicts of causal modeling; still others seem addicted to regression equations. We must be wary of treating any of these interval-analysis techniques as a panacea. Each technique can be useful when it is appropriate, and each can be misused by applying it mindlessly. The best uses of these techniques are guided by theory. The researcher who develops hypotheses based on a theory can legitimately employ these procedures to test those hypotheses. No single analysis procedure will ever be the answer to all data-analysis problems. Instead, the data analyst must be prepared to employ a range of analysis procedures to explore the data fully.

Questions

1. A correlation of .30 means that the independent variable explains __ percent of the variance in the dependent variable.
2. A correlation of .10 is significant in a large survey. That means:
   a. the relationship between these variables is important enough to study further.
   b. the true correlation is within .10 of the sample correlation.
   c. 10 percent of the variance is explained by the relationship.
   d. even if the relationship is too small to be very important, the relationship between the variables in the sample is more than sampling error can explain.
   e. none of the above.
3. The regression coefficient (b) in a single-predictor regression equation is found to equal .50. That means:
   a. the independent variable explains 25 percent of the variance in the dependent variable.
   b. a one-unit change in the independent variable induces a one-half-unit change in the dependent variable when the effects of other variables are controlled.
   c. The relationship is statistically significant.
   d. the correlation between the two variables is less than 1.0.
   e. none of the above.
4. Which variable is more important in the regression equation

$$Y' = 1200 + 15 \text{ (Years of education)} - 9 \text{ (Years of age)}?$$

12

Relationships Between Nominal and Interval Variables

Often we are interested in comparing two or more groups. The purpose might be to see if the groups are different with respect to some dependent variable or to see what effects the variable that creates the groups has on some dependent variable. For example, we might be interested in knowing whether males and females have different mean incomes. Alternatively stated, does gender affect mean income? The basic process is to determine whether some nominal-level independent variable (gender, in our example) affects some interval-level dependent variable (income).

There are a wide variety of techniques and ways of looking at such relationships, including both measures of association and tests of statistical significance. Three related approaches are described in this chapter: (*a*) testing whether the difference between two means is statistically significant, (*b*) measuring the proportion of the variance in the dependent variable that is accounted for by the nominal or categorical independent variable, and (*c*) testing whether the between-group differences are significantly greater than the within-group differences.

COMPARING TWO MEANS

To illustrate the nature of a grouping problem, say that we are studying county tax rates within a state and wish to determine whether the counties in the southern part of the state tend to have different tax rates from those in the northern part of the state. The data here are county-level data, and the grouping is into south and north.

Say that we have taken a simple random sample of eighteen counties in the state, have divided them into southern and northern counties, and that their tax rates are as shown in Table 12.1. The average tax rate for the southern counties is 20, while that for the northern counties is 25, so there is a difference of 5 in the average tax rates for the two sections. The important statistical question is whether that difference could be due to chance. Might the difference of 5 be due to the fact that we have examined only a sample of counties rather than examining the entire population of counties? Or, is the difference of 5 statistically significant? A difference-of-means test is required to answer these questions.

An equivalent way to think about this problem is to view section (south or north) as the independent variable and tax rate as the dependent variable. The dependent variable is measured at the interval level, but the independent variable is just a set of nominal categories. The statistical question described above amounts to asking

Table 12.1 Tax Rate by Section

Southern	Northern
10	15
9	14
11	16
20	25
19	24
21	26
30	35
29	34
31	36
$\bar{Y}_s = 20.00$	$\bar{Y}_n = 25.00$
$SS_s = 606$[a]	$SS_n = 606$[a]
$s_s^2 = 67.33$	$s_n^2 = 67.33$
$s_s = 8.21$	$s_n = 8.21$

$\bar{X}_t = 22.50$
$SS_t = 1324.5$
$s_t^2 = 73.58$
$s_t = 8.58$

[a]These measures will be explained later.
SOURCE: Hypothetical.

whether the nominal variable has a significant effect on the interval variable. However, the statistical techniques described in previous chapters do not cover this case of nominal independent variables and interval dependent variables.[1]

Critical Ratio

To determine how the statistical significance of the difference between two means should be tested, recall how the significance of a single mean was tested in Chapter 10. A critical ratio was computed, where the ratio was the difference between the observed and hypothesized means divided by the estimated standard error of the mean: $(\overline{Y} - \mu)/est\ s_m$. The test for the significance of a difference of means is similar in form. A critical ratio is taken. The numerator is the observed difference of means, instead of the difference between an observed mean and its hypothesized value. The denominator must also be an estimated standard error of the difference: $(est\ s_{diff})$. Thus, the critical ratio is $(\overline{Y}_1 - \overline{Y}_2)/(est\ s_{diff})$.

What is an "estimated standard error of the difference" and how is it computed? Conceptually, it should be clear that the significance of the observed difference of means depends on the dependent variable's variance. In Table 12.1, for example, the observed difference of means is $25 - 20 = 5$. Whether a difference of 5 is large depends on how much variance there is in the tax rates. If the tax rates have little variance, then 5 could be a large difference. If the tax rates have great variance, then 5 could be a small difference. More precisely, the significance of the observed difference of means depends on the dependent variable's variance within each category. If there is little variance in tax rates among southern counties and little variance in tax rates among northern counties, then 5 could be a large difference. If the tax rates in either (or both) sections have great variance, then 5 could be a small difference. The difference between sections is considered large only if it is large with respect to the differences within sections.

As was shown in Chapter 10, statistical testing requires knowing something about the underlying sampling distribution of the statistic

[1]Actually, the interval-level procedures described in the previous chapter could be used by treating the nominal independent variable as a dummy variable, but the new procedure described in this section is more straightforward to understand.

to be tested. In this case, we need to know about the sampling distribution of the difference between two means. The sample represented in Table 12.1 gives the difference of means for one possible sample. A different sample would probably yield a different difference of means. If we took a very large number of samples of the same size of the same population and examined the difference of means from each sample, we would have a sampling distribution for the difference between two means. If the sample size were large, the distribution would be normal. If not, it would be approximated by the t-distribution. In any case, it is the standard deviation of that distribution that is useful for inferential statistics, and it is called the *standard error of the difference of means.*

If we have separate independent samples from two groups whose populations variances are unknown and we assume to be unequal, then the standard error of the difference is:

$$est\ s_{\text{diff}} = \sqrt{s_{m_1}^2 + s_{m_2}^2}$$

or, using our formula for the standard error of the mean;

$$est\ s_{\text{diff}} = \sqrt{\frac{s_1^2}{N_1 - 1} + \frac{s_2^2}{N_2 - 1}}$$

If we make the additional assumption that the variances of the two populations are equal but still unknown, we can use both sample standard deviations to estimate the one underlying population standard deviation.

$$est\ s_{\text{diff}} = \sqrt{\frac{N_1 s_1^2 + N_2 s_2^2}{N_1 + N_2 - 2}}\sqrt{\frac{1}{N_1} + \frac{1}{N_2}}$$

$$= \sqrt{\frac{\sum_{i=1}^{N_1} (Y_{1i} - \overline{Y}_1)^2 + \sum_{i=1}^{N_2} (Y_{2i} - \overline{Y}_2)^2}{N_1 + N_2 - 2}}\sqrt{\frac{1}{N_1} + \frac{1}{N_2}}$$

The first term is the combined (or "pooled") estimate of the standard deviation of the dependent variable.

Returning to the example, the pooled estimate of the standard deviation is:

$$\sqrt{\frac{9(8.21)^2 + 9(8.21)^2}{9 + 9 - 2}} = \sqrt{\frac{606 + 606}{16}} = \sqrt{\frac{1212}{16}} = 8.70$$

Therefore, the estimated standard error of the difference is:

$$est\ s_{\text{diff}} = 8.70\sqrt{\frac{1}{9}+\frac{1}{9}} = 8.70\ (.47) = 4.10$$

The critical ratio for a 5-unit difference is then $5/4.10 = 1.22$. To determine whether this is statistically significant, we employ a t-test using $(N_1 + N_2 - 2) = 16$ degrees of freedom. According to Table 10.3, a critical ratio of at least 2.12 would be required for statistical significance at the .05 level with 16 degrees of freedom, so the difference is not significant at the .05 level. That is, there is more than a 5 percent chance of obtaining the observed difference by chance in our sample if the two sections had the same average tax rate. Therefore, we would conclude that the data do not demonstrate that the two sections have significantly different tax rates.

Confidence Interval

It is also possible to construct confidence intervals for the difference between two means. The confidence interval is the observed difference $\pm t\ (est\ s_{\text{diff}})$, where the estimated standard error of the difference is computed as above.

For the example of Table 12.1, there are 18 cases, so the number of degrees of freedom for the t value would be $18 - 2 = 16$. According to Table 10.3, a t value of 2.12 corresponds to a 95 percent confidence interval with 16 degrees of freedom. Because the estimated standard error of the difference was found to be 4.10, the confidence interval is the observed difference, 5, $\pm$ 2.12 (4.10), or 5 $\pm$ 8.69, or -3.69 to 13.69.

In other words, given our sample sizes we would expect the sectional means to differ by as much as 8.69 due to random fluctuations in the sample 95 percent of the time. We would conclude, therefore, that the means in Table 12.1 could differ as much as they do by chance alone. Another way to say this is to note that the confidence interval includes zero, so we cannot be sure that the two population means really differ from one another.

The confidence interval tells you the maximum amount of difference that you might expect due to chance. Generally speaking, differences larger than that are statistically significant, and smaller ones are not. An observed difference less than t times $(est\ s_{\text{diff}})$ could represent no actual difference between the two population means.

Thus, in the present example, the observed difference in means would have to be larger than 8.69 (or smaller than -8.69) to be statistically significant.

Assumptions

The t distribution has been used so far in this section because of the small sample sizes. For large samples, the normal distribution would be used.

Also, if the population variance is known, then the population standard deviation would be substituted for the pooled standard deviation in the formula for the standard error of the difference.[2] Where population variances are both known and different:

$$s_{\text{diff}} = \sqrt{\frac{s_1^2}{N_1} + \frac{s_2^2}{N_2}}$$

Where population variances are equal and known:

$$s_{\text{diff}} = s\sqrt{\frac{1}{N_1} + \frac{1}{N_2}}$$

Even using the t-distribution, it is necessary that the populations sampled be normal and that the population variances be nearly equal. Fortunately, these assumptions can be violated without harm if the sample sizes are large (for the normality assumption) and of equal size (for the equal variance assumption).

Finally, it would be fallacious to do a large number of t-tests (or confidence intervals) at the .05 level and report only those that are statistically significant. After all, 5 percent of the tests would be significant by chance alone. Actually even more than 5 percent would be significant by chance if the tests were not independent of one another, as if we separately tested the significance of the difference in mean tax rates between northern and central counties, central and southern counties, and northern and southern counties. Different statistical procedures, such as those described later in this chapter, must be used when comparing several means.

---

[2]Actually, it is not quite that simple; the $N - 1$ in the denominator should be replaced by N.

MEASURING STRENGTH OF RELATIONSHIP

In addition to testing whether the difference between two means is statistically significant, we can examine the extent to which the categorical independent variable is related to the interval dependent variable. We will develop a measure of the strength of the relationship that will show how much of the variance in the dependent variable can be accounted for by the independent variable.

Extreme Relationships

A useful beginning point is to consider what would be a perfect relationship between a nominal and an interval variable. If we think of a relationship in terms of predictability (see Chapter 7), a perfect relationship would mean that we could predict the interval dependent variable value exactly from the person's category on the independent variable. Say we observed variance in people's heights and wanted to explain that variance in terms of their gender. Table 12.2 illustrates what a perfect relationship between the nominal independent variable (sex) and the interval-level dependent variable (height) would look like. All men, in this example, have a height of 72 inches, while all women have a height of 62 inches, so we could perfectly predict people's heights from their genders. This is a perfect relationship, since the variance (for the variable height) within the two gender groups is zero.

By contrast, what would constitute no relationship between the nominal and interval variables? There would be no relationship if the dependent variable means were identical for each category of the independent variable, because knowing the independent variable

Table 12.2 Height by Sex

	Males	Females
	72	62
	72	62
	72	62
	72	62
Means:	$\overline{Y}_m = 72$	$\overline{Y}_f = 62$
Variances:	$s_m = 0$	$s_f = 0$
	Grand mean = 65	
	Total variance = 25	

SOURCE: Hypothetical.

Table 12.3 Age by Sex

	Males	Females
	16	21
	18	12
	16	18
	22	21
Means:	$\overline{Y}_m = 18.0$	$\overline{Y}_f = 18.0$
Variances:	$s_m = 6.0$	$s_f = 13.5$
	Grand mean = 18.0	
	Total variance = 9.75	

SOURCE: Hypothetical.

would then not help us predict the dependent variable value. Say we observed variance in people's ages and wanted to explain that variance in terms of their genders. Table 12.3 illustrates what no relationship between the nominal independent variable (sex) and the interval-level dependent variable (age) would look like. The men and women have identical mean ages of eighteen. Our best guess would be that a man is age eighteen and that a woman is age eighteen, which would also be our best guess if we did not know the person's sex.

As these opposite examples illustrate, the extent of predictability between a categoric variable and an interval variable depends on how much the group means differ and on how little variance there is within the categories of the independent variable. A strong relationship occurs when the group means differ considerably and when there is little variance remaining within any group; a weak relationship occurs when the group means are similar and when there is considerable variance within all the groups.

A Measure of Predictability

The same logic used to develop lambda and r^2 in previous chapters can be used to derive a measure of predictive association between a nominal variable and an interval dependent variable. The measure is called "eta-squared" (η^2), or the correlation ratio (η is the Greek letter *eta*). Eta-squared has precisely the same interpretation as r^2, namely the proportion of variance in the dependent variable attributable to (or explained by) the independent variable. Equivalently, it can also be interpreted, like lambda, as the proportion of reduction in error in guessing the value of the dependent variable by knowing the value of the independent variable.

If eta-squared is to be analogous to r^2, the total amount of variance in the dependent variable (s_Y^2) must be calculated, along with either the amount of variance "explained" by the independent variable or the amount of variance "unexplained" by the independent variable (also known as the "remaining" or "residual" variance, $s_{Y'}^2$). Then, applying the r^2 formula from Chapter 11 gives:

$$\eta^2 = r^2 = \frac{\text{explained variance}}{\text{total variance}}$$

$$= \frac{\text{total variance} - \text{unexplained variance}}{\text{total variance}} = \frac{s_Y^2 - s_{Y'}^2}{s_Y^2}$$

The unexplained variance is calculated in the same way as the pooled variance in the previous section—weighting the variance in each group by the number of cases in that group and then dividing by the total number of cases. Returning to Table 12.3, for example, the variance for the four men is 6 and the variance for the four women is 13.5, so the unexplained variance is $(4[6] + 4[13.5])/8 = (24 + 54)/8 = 78/8 = 9.75$. Applying the above formula for Table 12.3 then gives an eta-squared of $(9.75 - 9.75)/9.75 = 0.0$.

We can now return to the tax-rate example in the previous section. The total variance of the tax rates was 73.58. The variance within the southern counties was 67.33 and the variance within the northern counties was 67.33. Therefore the unexplained variance was $(9[67.33] + 9[67.33])/18 = (606 + 606)/18 = 1{,}212/18 = 67.33$. The extent of predictability is then:

$$\eta^2 = \frac{73.58 - 67.33}{73.58} = \frac{6.25}{73.58} = .085$$

That is, whether the county is located in the north part of the state or the south accounts for 8.5 percent of the variance in the tax rates of the counties. The significance test in the previous section demonstrated that this relationship is not statistically significant.

Computational Method

We have based the eta-squared measure so far on the prior calculation of the total and unexplained variances. However, this is not necessary, and the procedure is rarely done this way. Recall that the total variance for a population is calculated by taking the deviations of each case from the mean, squaring those deviations, summing the squared deviations, and dividing by the number of cases:

$$s_Y^2 = \frac{(Y_i - \overline{Y})^2}{N}$$

The unexplained variance is also based on averaging a set of squared deviations. As in Chapter 11, the deviations are the difference between the observed value and predicted value. In this case, the predicted value is the appropriate group mean.

$$s_{Y'}^2 = \frac{\sum_j \sum_i (Y_{ij} - \overline{Y}_j)^2}{N}$$

Because each variance in the eta-squared formula involves dividing a sum of squares by N, the total number of cases, the N term cancels out. Eta-squared can then be written solely in terms of sums of squares.

Let SST represent the total sum of squared deviations, $\Sigma(Y_i - \overline{Y})^2$, to be called the "total sum of squares." Let SSB represent the between-groups sum of squares, which corresponds to the explained variance and represents the dispersion of the group means from the grand mean, $\Sigma N_j(\overline{Y}_j - \overline{Y})^2$. Finally, let SSW represent the within-groups sum of squares, $\Sigma\Sigma(Y_{ij} - \overline{Y}_j)^2$, which corresponds to the unexplained variance. Then the correlation ratio is:

$$\eta^2 = \frac{SSB}{SST} = \frac{SST - SSW}{SST}$$

SST is the total variance, before dividing by N. SSW is the sum of the sum of squares within each group.

Recalculating the tax-rate example, the total sum of squares was 1324.5. The sum of squares within the southern group was 606, and the sum of squares within the northern group was 606, so the within-groups sum of squares is 1,212. Therefore,

$$\eta^2 = \frac{SST - SSW}{SST} = \frac{1{,}324.5 - 1{,}212}{1{,}324.5} = \frac{112.5}{1{,}324.5} = .085$$

Multicategory Variables

The value of eta-squared in the example is the same as the r^2 value would be if we coded southern as "0" and northern as "1" and correlated section with the tax rate. Of course, this identity holds only because the independent variable is dichotomous. Eta-squared is useful precisely because it applies equally for multicategory independent variables.

Instead of dividing the eighteen counties according to section or north-south, say that we separate them into three categories according to topography, as shown in Table 12.4. The group means are more distinct from one another than they were for the sectional categorization. And the sums of squares within the topography groups are smaller than they were for the sectional classification. The topography groups have sharply different mean tax rates and are fairly homogeneous internally.

To calculate the eta-squared, we must first compute the sum of squares total, which is 1,324.5, as it was for the original data. Also, we must compute the sum of squares within by summing the sum of squares within each category: 41.5 + 41.5 + 41.5 = 124.5. Then the correlation ratio is

$$\eta^2 = \frac{1{,}324.5 - 124.5}{1{,}324.5} = \frac{1{,}200}{1{,}324.5} = .906$$

In words, dividing the counties into three groups based on their topography allowed us to explain 90.6 percent of the variation in their tax rates. Alternatively, there is a 90.6 percent reduction in error in estimating the tax rate of a county by knowing if it is the mountains, near the coast, or in the valley.

Table 12.4 County Tax Rates by Topography

	Mountain	Valley	Coastal
	10	19	31
	14	25	29
	9	26	36
	11	21	34
	16	20	35
	15	24	30
Means:	$Y_m = 12.5$	$Y_v = 22.5$	$Y_c = 32.5$
Sum of Squares:	41.5	41.5	41.5

Grand mean = 22.5
Sum of squares (total) = 1,324.5
Sum of squares within = 41.5 + 41.5 + 41.5 = 124.5

SOURCE: Hypothetical.

Clearly, whether a county is coastal or valley or mountain is much more important in determining its tax rate than whether it is southern or northern. The difference of means at the beginning of this chapter showed that the sectional distinction was not statistically significant. But what about the topographic classification? Is the 90.6 percent reduction in error statistically significant? We could try to determine its significance by applying the difference-of-means test, but that test was only for two groups, not three, as here. We could apply the difference-of-means test first to test whether the difference between the mountain and coastal counties is significant, and then to test whether the difference between the valley and coastal counties is significant, and finally to test whether the difference between mountain and valley counties is significant. However, that testing procedure would be fallacious, because the three tests of significance would not be independent of one another. What we require is a single test of whether the topographic distinction is statistically significant. A statistical technique that provides such a test is *analysis of variance.*

ONE-WAY ANALYSIS OF VARIANCE

Analysis of variance provides us with a means of assessing the statistical significance of the relationship between nominal- and interval-level variables. We begin with "one-way" analysis of variance, which is used when there is a single nominal independent variable. Later in this chapter we turn to "two-way" analysis of variance, which is appropriate when there are two independent variables, the nominal analogue of multiple-regression analysis.

Logic of Comparing Variances

One-way analysis of variance allows us to ascertain the probability that the relationship found for a sample holds for the underlying population. The principle on which this technique is based is that there are three possible estimates of the population variance that could be made from the data. First, the within-group variance, weighted by group size, and analogous to the pooled variance in the difference of means tests, could be used. Second, the between-groups variance could be used as an estimator. It is the deviation of each group mean ($\overline{Y}_i$) from the grand mean ($\overline{Y}$). The third estimator of the population variance would be simply to calculate the sample variance without dividing the data into groups. (It would be the sum of the squared deviations of each case from the grand mean divided by $N - 1$).

While it may appear that we have three estimators of the population variance, they are all related to one another such that any one is a linear combination of the other two. (Total variance = within-groups variance + between-groups variance.) Therefore, we have but two independent estimators.

Analysis of variance uses the between-groups and the within-groups variances as the estimators. The extent to which these two estimators differ will indicate the likelihood that the relationship is due to chance. The within-groups variance could be the same whether the group means are really different in the population or not. It would therefore accurately estimate the population variance. The between-groups variance, however, is only an accurate estimate of the population variance if the true group means are the same. If they are not the same, the between-groups variance will overestimate the population variance. Consequently, if we compute the ratio of the between-groups variance to the within-groups variance, we should get a number close to 1 if there is no relationship between the independent and dependent variables in the population. If this ratio, called the "F ratio" or simply F, is much larger than 1, then it is likely that the relationship is true of the population.

This F value is calculated for an analysis of variance and compared to standard values of F calculated from the known distribution of F. The calculated value of F must be larger than the value of F found in the table for the appropriate level of statistical significance and numbers of degrees of freedom.

Estimates of the population variance (called "mean squares") are calculated from the proper sum of squares by dividing by the appropriate number of "degrees of freedom." The total number of degrees of freedom for a sample is one less than the number of observations: $N - 1$. The sum of the number of degrees of freedom for the between-groups variance and the number of degrees of freedom for the within-groups variance must equal the total number of degrees of freedom:

$$\text{total } df = \text{between } df + \text{within } df$$

The between-groups degrees of freedom is one less than the number of groups: $k - 1$, where k is the number of categories of the independent variable. The number of degrees of freedom for the within-groups variance is the number of observations minus the number of groups: $N - k$. As stated above, the between-groups degrees of freedom and the within-groups degrees of freedom sum up to the total degrees of freedom:

$$N - 1 = (k - 1) + (N - k)$$

Analysis-of-Variance Table

While the jargon may now seem overpowering, the analysis-of-variance calculations are quite simple. Table 12.5 is the "analysis-of-variance-table" for the data in Table 12.4. Reading the first column from the bottom up, the total sum of squares is 1,324.5, as stated previously. The within-groups sum of squares is 124.5, as calculated in the previous section. The total sum of squares is just the sum of the within-groups sum of squares and the between-groups sum of squares, so the between-groups sum of squares is the difference between the sum of squares total (1,324.5) and the sum of squares within (124.5), or 1,200. The second column shows the corresponding numbers of degrees of freedom. There were 18 observations, so the total number of degrees of freedom is 17. There were three groups, so the between-groups degrees of freedom is $3 - 1 = 2$. The within-groups degrees of freedom is $18 - 3$ (or, if you prefer, 17 total degrees of freedom minus 2 degrees of freedom between) $= 15$.

Table 12.5 Analysis-of-Variance Table: County Tax Rate by Topography

Source of Variation	Sum of Squares	*df*	Mean Square (s)	*F* Ratio	Required *F* Ratio	Significant?
Between-Groups	1,200	2	600	72.3	3.68	Yes
Within-Groups	124.5	15	8.3			
Total	1,324.5	17				

$$F = \frac{\hat{s}_b^2}{\hat{s}_w^2} = \frac{MSB}{MSW} = \frac{600}{8.3} = 72.3$$

$$\eta^2 = \frac{SSB}{SST} = \frac{1{,}200}{1{,}324.5} = .906$$

The calculations of the estimated population variance ($\hat{s}^2$) are then made by dividing the sum of squares for each type of variance by the number of degrees of freedom for that type. For example, the between-groups mean square is the sum of squares between (1,200) divided by its degrees of freedom (2) to give 600. Similarly, the within-groups mean square is the sum of squares within (124.5) divided by its degrees of freedom (15) to give 8.3.

Table 12.6 Distribution of F (p = .05)

df_{den} \ df_{num}	1	2	3	4	5	6	8	12	24	∞
1	161.4	199.5	215.7	224.6	230.2	234.0	238.9	243.9	249.0	254.3
2	18.51	19.00	19.16	19.25	19.30	19.33	19.37	19.41	19.45	19.50
3	10.13	9.55	9.28	9.12	9.01	8.94	8.84	8.74	8.64	8.53
4	7.71	6.94	6.59	6.39	6.26	6.16	6.04	5.91	5.77	5.63
5	6.61	5.79	5.41	5.19	5.05	4.95	4.82	4.68	4.53	4.36
6	5.99	5.14	4.76	4.53	4.39	4.28	4.15	4.00	3.84	3.67
7	5.59	4.74	4.35	4.12	3.97	3.87	3.73	3.57	3.41	3.23
8	5.32	4.46	4.07	3.84	3.69	3.58	3.44	3.28	3.12	2.93
9	5.12	4.26	3.86	3.63	3.48	3.37	3.23	3.07	2.90	2.71
10	4.96	4.10	3.71	3.48	3.33	3.22	3.07	2.91	2.74	2.54
11	4.84	3.98	3.59	3.36	3.20	3.09	2.95	2.79	2.61	2.40
12	4.75	3.88	3.49	3.26	3.11	3.00	2.85	2.69	2.50	2.30
13	4.67	3.80	3.41	3.18	3.02	2.92	2.77	2.60	2.42	2.21
14	4.60	3.74	3.34	3.11	2.96	2.85	2.70	2.53	2.35	2.13
15	4.54	3.68	3.29	3.06	2.90	2.79	2.64	2.48	2.29	2.07
16	4.49	3.63	3.24	3.01	2.85	2.74	2.59	2.42	2.24	2.01
17	4.45	3.59	3.20	2.96	2.81	2.70	2.55	2.38	2.19	1.96
18	4.41	3.55	3.16	2.93	2.77	2.66	2.51	2.34	2.15	1.92
19	4.38	3.52	3.13	2.90	2.74	2.63	2.48	2.31	2.11	1.88
20	4.35	3.49	3.10	2.87	2.71	2.60	2.45	2.28	2.08	1.84
21	4.32	3.47	3.07	2.84	2.68	2.57	2.42	2.25	2.05	1.81
22	4.30	3.44	3.05	2.82	2.66	2.55	2.40	2.23	2.03	1.78
23	4.28	3.42	3.03	2.80	2.64	2.53	2.38	2.20	2.00	1.76
24	4.26	3.40	3.01	2.78	2.62	2.51	2.36	2.18	1.98	1.73
25	4.24	3.38	2.99	2.76	2.60	2.49	2.34	2.16	1.96	1.71
26	4.22	3.37	2.98	2.74	2.59	2.47	2.32	2.15	1.95	1.69
27	4.21	3.35	2.96	2.73	2.57	2.46	2.30	2.13	1.93	1.67
28	4.20	3.34	2.95	2.71	2.56	2.44	2.29	2.12	1.91	1.65
29	4.18	3.33	2.93	2.70	2.54	2.43	2.28	2.10	1.90	1.64
30	4.17	3.32	2.92	2.69	2.53	2.42	2.27	2.09	1.89	1.62
40	4.08	3.23	2.84	2.61	2.45	2.34	2.18	2.00	1.79	1.51
60	4.00	3.15	2.76	2.52	2.37	2.26	2.10	1.92	1.70	1.39
120	3.92	3.07	2.68	2.45	2.29	2.17	2.02	1.83	1.61	1.25
∞	3.84	2.99	2.60	2.37	2.21	2.09	1.94	1.75	1.52	1.00

SOURCE: Abridged from R. A. Fisher and F. Yates, *Statistical Tables for Biological, Agricultural, and Medical Research* (Edinburgh and London: Oliver & Boyd, 1948), Table V, by permission of the authors and publishers.

The F ratio, or F, can be calculated from these two estimates. F is the ratio of explained to unexplained variance, the between-groups mean square divided by the within-groups mean square. In this example, it is $600/8.3 = 72.3$. In order to complete the significance test, we must compare the F ratio we calculated with that which we might expect to get by chance when randomly allocating eighteen observations to three categories.

Table 12.6 is the table of the F distribution for the .05 level of significance. The entries in the table are based on the two different degrees of freedom, the one for the numerator of the F ratio (here the between-groups degrees of freedom) and the one for the denominator (the within-groups degrees of freedom). The columns of the table correspond to the numerator of the F ratio and the rows to the denominator, so look in column 2 (for 2 degrees of freedom between) and row 15 (for 15 degrees of freedom within), where you should find a value of 3.68. If our obtained F ratio is greater than or equal to 3.68, then there would be less than a 5 percent chance that the relationship is statistically significant at the .05 level. Here 72.3 is larger than 3.68, so we conclude the relationship is significant. If the obtained F ratio were less than 3.68, we would conclude that the relationship was not statistically significant at the .05 level. Tables of the F distribution also exist for other significance levels.

Alternative Computational Procedure

When we have already calculated eta-squared, a simple way to calculate F is from that value:

$$F = \left(\frac{\eta^2}{1 - \eta^2}\right)\left(\frac{N - k}{k - 1}\right)$$

In our tax-rate example, eta-squared was .906, so F would be:

$$F = \left(\frac{.906}{1 - .906}\right)\left(\frac{18 - 3}{3 - 1}\right) = (9.64)(7.5) = 72.3$$

Another Example

As another example, we can return to our original data showing county tax rates by section. The eta-squared value was .085, for eighteen observations (N) and two groups (k), so F can be computed as:

$$F = \left(\frac{.085}{1 - .085}\right)\left(\frac{18 - 2}{2 - 1}\right) = (.09)(16) = 1.49$$

To test the significance of this F ratio, we would use Table 12.6 again, with 1 degree of freedom for the numerator (two groups minus one) and 16 degrees of freedom for the denominator (eighteen observations minus two groups). In column 1, row 16, of Table 12.6, the F value is 4.49. The obtained F ratio is less than the F value required for significance at the .05 level, so we would conclude that the difference between the southern and northern counties in their tax rates is not statistically significant.

This example is actually the same example as we originally used for the difference-of-means test at the beginning of this chapter. There we used a significance test based on the t-distribution. Here we use a significance test based on the F-distribution. One might be worried that these are two different tests that could yield different results for the same data. In fact, however, there is no such difficulty. When there are two groups, F values are just the square of the corresponding t values. The t value we obtained for these data at the beginning of this chapter was 1.22, and the F of 1.49 is just $(1.22)^2$. The t value required for significance in the difference-of-means test was 2.12, and the F of 4.49 is just $(2.12)^2$. Because of this correspondence, the F-test will be significant if and only if the t-test was significant. These are fully equivalent tests, except that the F-test generalizes to permit more than two independent variable categories.

Table 12.7 shows the full analysis-of-variance table for this example. The total sum of squares is 1,324.5, and the total degrees of freedom is

Table 12.7 Analysis-of-Variance Table: County Tax Rate by Section

Source of Variation	Sum of Squares	*df*	Mean Square (s)	F Ratio	Required F Ratio	Significant?
Between-Groups	112.5	1	112.5	1.49	4.49	No
Within-Groups	1,212.0	16	75.75			
Total	1,324.5	17				

$$F = \frac{\hat{s}_b^2}{\hat{s}_w^2} = \frac{MSB}{MSW} = \frac{112.5}{75.75} = 1.49$$

$$\eta^2 = \frac{112.5}{1{,}324.5} = .085$$

seventeen, as in Table 12.5. The within-groups sum of squares is now 606 + 606 = 1,212, with sixteen degrees of freedom, for a mean square of 75.75. The between-groups sum of squares is the difference between the total (1,324.5) and the within (1,212), or 112.5. The between-groups degrees of freedom is one less than the number of groups: 2 − 1 = 1. So the between-groups mean square is 112.5 divided by one, or 112.5. The F ratio is the between mean square (112.5) divided by the within mean square (75.75), or 1.49. It is conventional to summarize analysis of variance tests with the analysis-of-variance table as shown in Tables 12.5 and 12.7.

Assumptions

Several cautions are in order. The tests of significance are possible because of certain assumptions that are made about the data. First, it is assumed that the data are collected by simple random sampling. Second, the dependent variable is assumed to be normally distributed in each of the groups. Third, the population variances for each group must be assumed to be the same.

The appropriateness of the first assumption—random sampling—can easily be ascertained by checking the data-collection procedure. The normality assumption is often very difficult to check, but violations of it are not critical for large samples. The third assumption—equal population variances—cannot usually be checked directly, but an indication of its validity can be obtained by comparing the variances of the groups in the sample. If they are relatively similar, then the assumption of equal variances in the population is probably acceptable. This assumption is critical, but can be ignored when there are equal numbers of cases within each group.

TWO-WAY ANALYSIS OF VARIANCE

The analysis-of-variance procedure can be extended to more than one nominal independent variable just as the regression procedure of Chapter 11 was extended to multiple-interval independent variables. (There is also a mixed procedure for one nominal independent variable and one interval independent variable, known as *analysis of covariance,* but we shall not consider it here.) The procedure for two nominal independent variables is known as "two-way" analysis of variance.

Design Considerations

The two-way analysis-of-variance procedure depends directly on the research design. There are several different models for different possible designs.

One basic issue is how the independent variables are related to one another. A "nested" model is one in which the independent variables have a hierarchical relationship to one another. Say, for example, we were comparing the voting turnout rates by congressional districts within states, with separate observations for each of the five elections in a decade.[3] We would be interested in determining how much of the variation in turnout rates is due to state differences and how much of the variation is due to differences between congressional districts. However, we could not compare congressional districts across states; they exist only within the states. It would be silly to compare the voting turnout across states in congressional district number 1 versus congressional district number 2, because the district numbers are arbitrary and have no interstate meaning.

By contrast, a "completely crossed" model is one in which the dependent variable is measured for each combination of independent variables. Our county tax rate example would be completely crossed if there are mountain, valley, and coastal communities within both north and south sections. We shall concentrate in this book on the completely crossed design.

A second basic issue is how the categories of the variables are chosen. In a "fixed-effects" model, inferences are to be made only for the actual categories. In a "random-effects" model, the categories analyzed are a random sample of the possible categories and the inferences are to be made to the total set of possible categories. For example, using men and women as categories of a gender variable would involve the fixed-effects model, while using thirty-year-olds versus forty-year-olds to study the effects of age would involve the

---

[3]This design is based on Donald E. Stokes, "A Variance Components Model of Political Effects," in John M. Claunch (ed.), *Mathematical Applications in Political Science,* Vol. 1 (Dallas: Arnold Foundation, 1965, 61–85).

random-effects model. Only the fixed-effects model can be considered here; the random-effects model is too complex for present purposes.

A third important design issue involves the number of cases with each combination of independent variables. The design is termed "balanced" if there are an equal number of observations for each combination of the independent variables, and "unbalanced" if there are different numbers of observations for different combinations of independent variables. In a balanced design, the two independent variables are uncorrelated. While there are procedures for analyzing the unbalanced case, the analysis of variance calculations are much simpler for the balanced design. We shall consider only the balanced case in this book.

Analysis Considerations

There are two advantages in conducting a two-way analysis of variance when there are two independent variables instead of just doing separate one-way analysis of variance for each independent variable.

First, not only the "separate effects" of the two independent variables can be studied but their combined effects can also be determined. In a completely crossed design, the effect of one independent variable may depend on the category of the other independent variable. In the tax rate example, the effect of the topography may be different in northern counties than in southern counties. This combination effect is known technically as an "interaction effect."

The model is termed "additive" if there are no interaction effects. Whether the data are additive or if there is an interaction effect is an empirical question. The significance of the interaction effect is tested with an F-test just as that of each separate effect is tested. We begin this section with the additive case, but will illustrate the interactive case later.

Second, the two-way analysis-of-variance procedure is important because a variable may be found to have significant separate effects when the other variable's effects are controlled even if it did not have separate effects in a one-way analysis of variance. Consider the tax-rate example again. The sectional distinction was not significant in the one-way analysis of variance. However, the topography classification was so important that it would have been nearly impossible for any other variable to have a significant effect. Still, the sectional distinction might be significant *within* categories of the topography classification. Stated in a different way, the sectional eta-squared of .085 seems small when considered by itself, but it actually is a very large part of the variance that the topography eta-squared of .906 left unexplained. The sectional distinction could be significant once the topography effect is separated out. Two-way analysis of variance examines this possibility.

Additive Case

Logic of Comparing Variances. We now begin with the additive case for a fixed-effects model with a completely crossed and balanced design. If the jargon seems imposing, the analysis is actually very similar to that already explained in this chapter. Because we have assumed a completely crossed design, the between-groups variances are calculated exactly as for the one-way case, except that there are now two between-groups variances, one for each independent variable. Because we have assumed a balanced design, the within-groups sum of squares is the total sum of squares minus the sum of squares attributable to the first independent variable minus the sum of squares attributable to the second independent variable. Similarly, the within-groups degrees of freedom is the total number of degrees of freedom minus that for the first independent variable minus that for the second independent variable. The within mean square is the within-groups sum of squares divided by this degree of freedom value. The significance of each independent variable is ascertained by dividing its mean square between by the new within mean square and then comparing that F ratio with the F value required for the corresponding numbers of degrees of freedom.

Table 12.8 shows the tax rates of the counties by combinations of topography and sectional categories. Obviously, the northern coastal counties have much higher tax rates than the southern ones, and the same is true within the valley counties and within the mountain counties. The overall difference between the southern and northern counties may not be substantial, but there is a systematic effect evident once the topography classification is controlled. This will be evident in the analysis-of-variance results.

Table 12.8 County Tax Rates by Topography and Section

	Mountain	Valley	Coastal
Southern	10	20	29
	11	19	31
	9	21	30
Northern	15	24	36
	14	25	35
	16	26	34

SOURCE: Hypothetical.

Table 12.9 gives the two-way analysis-of-variance table for the tax-rate data. The "between-columns" line refers to the variance due to the differences among the column categories of Table 12.8, the differences among mountain, valley, and coast categories. The between-columns sum of squares is the same as the sum of squares between in the one-way analysis of variance using topography as the only predictor, namely 1,200; its degree of freedom is the number of columns minus one, $3 - 1 = 2$; and its mean square is its sum of squares divided by the degrees of freedom, $1{,}200/2 = 600$.

The "between-rows" line refers to the variance due to the differences between the row categories of Table 12.8, the differences between southern and northern categories. The between-rows sum of squares is the same as the sum of squares between in the one-way analysis of variance using section as the only predictor, namely 112.5; its degrees of freedom is the number of rows minus 1, $2 - 1 = 1$; and its mean square is its sum of squares divided by the degrees of freedom, $112.5/1 = 112.5$.

Table 12.9 Two-Way Analysis-of-Variance Table: Tax Rate by Topography and Section, Additive Case

Source of Variation	Sum of Squares	*df*	Main Square (*s*)	*F* Ratio	Required *F* Ratio	Significant?
Columns	1,200	2	600	700	3.74	Yes
Rows	112.5	1	112.5	131.25	4.60	Yes
Within	12	14	.86			
Total	1,324.5	17				

The total sum of squares and the total degrees of freedom are the same as in the one-way analysis of variance. The within-cells sum of squares is now the total sum of squares (1,324.5) minus the between-columns sum of squares (1,200) minus the between-rows sum of squares (112.5), to leave a remainder of 12. The within-cells degrees of freedom is the total degrees of freedom (17) minus the between-columns value (2) minus the between-rows value (1), to leave a remainder of 14. The within-cells mean square is the within sum of squares (12) divided by its degrees of freedom (14), to give .86.

The between-columns F ratio is the between-columns mean square (600) divided by the within mean square (.86), or 700. With two degrees of freedom for the numerator of this ratio and fourteen for the denominator, an F value of 3.74 is required for significance at the .05 level (column 2, row 14, of Table 12.6). Therefore, the topography variable is judged to have a significant effect on tax rate, as it was in the one-way analysis of variance.

The between-rows F ratio is the between-rows mean square (112.5) divided by the within mean square (.86), or 131.25. With one degree of freedom for the numerator of this ratio and fourteen for the denominator, an F value of 4.60 is required for significance at the .05 level (column 1, row 14, of Table 12.6). Therefore, the section variable is also judged to have a significant effect on tax rate, an effect that did not appear in the one-way analysis of variance because the effect of topography classification had not been first removed.

Additivity. The analysis of tax-rate data has assumed that the two independent variables do not have combination effects. We can examine this assumption by computing the mean tax rate for each combination of the two variables, as shown in Table 12.10. Notice that the values in the second column are always 10 points above those in the first column and always 10 points below those in the third

Table 12.10 Mean Tax Rates by Topography and Section

	Mountain	Valley	Coastal	Mean	Effect
Southern	10	20	30	20	−2.5
Northern	15	25	35	25	+2.5
Mean	12.5	22.5	32.5	22.5	
Effect	−10	0	+10		

Table 12.11 Expected Mean Tax Rates Under Additivity

	Mountain	Valley	Coastal	Mean	Effect
	22.5	22.5	22.5		
	−2.5	−2.5	−2.5		
Southern	−10.0	+0.0	+10.0		
	10.0	20.0	30.0	20	−2.5
	22.5	22.5	22.5		
	+2.5	+2.5	+2.5		
Northern	−10.0	+0.0	+10.0		
	15.0	25.0	35.0	25	+2.5
Mean	12.5	22.5	32.5	22.5	
Effect	−10.0	0.0	+10.0		

column. (Equivalently for our purposes, the values in the second row are always 5 points above those in the first row.) This regularity shows that there are not unique effects due to the combinations of the two independent variables.

The fourth column of Table 12.10 shows that the mean tax rate for all southern counties is 20, while that for all northern counties is 25. Since the overall mean is 22.5, the "effect" associated with being a southern county is $20 - 22.5 = -2.5$ and the effect associated with being a northern county is $25 - 22.5 = +2.5$ (see last column of Table 12.10).

The third row of Table 12.10 shows that the mean tax rate for the mountain counties is 12.5, for the valley counties is 22.5, and for the coastal counties is 32.5. A calculation similar to that above gives an effect of -10 for mountain ($12.5 - 22.5$), 0 for valley ($22.5 - 22.5$), and $+10$ for coastal ($32.5 - 22.5$), as shown in the bottom row of the table.

Now, if the data are strictly additive, the mean value for every cell should equal the grand mean plus the effects of that cell's row plus the effect of that cell's column. For example, in an additive situation, the mean value for the southern mountain cell would be the grand mean (22.5) plus the effect of the southern row (-2.5) plus the effect of the mountain column (-10), $22.5 - 2.5 - 10 = 10$. Table 12.11 shows the calculation of the cell means expected under additivity for all of the cells. Because the actual cell means of Table 12.10 are all identical to those expected under additivity according to Table 12.11, the effects of topography and section are completely additive. There is no interaction effect present in these data.

Incidentally, there is a simple calculation procedure for the columns sum of squares and rows sum of squares once the effect of each column and of each row have been calculated as in Table 12.10. The columns sum of square can be computed by squaring the effect of each column, summing those squares, and then multiplying that sum by the number of observations in each category. In Table 12.10, this involves adding the squares of the effects of the mountain column (-10^2), the valley column (0^2), and the coastal column ($+10^2$) to get a total of 200 and multiplying the number of cases per column (6) to get the between-columns sum of squares of 1,200. Similarly, the between-rows sum of squares is obtained by squaring the effect of the first row ($-2.5^2 = 6.25$), squaring the effect of the second row ($+2.5^2 = 6.25$), summing the squares ($6.25 + 6.25 = 12.5$), and multiplying by the number of cases in each row ($12.5 \times 9 = 112.5$). This procedure always works for the completely crossed and balanced design.

Interactive Case

To illustrate the interactive case, say that the county tax-rate data were as in Table 12.12 instead of as in Table 12.8. The means for each row and for each column are unchanged, and so the sums of squares for section and topography are unchanged. However, the data are no longer additive. That can be seen by inspecting the cell means, as shown in Table 12.13. Overall, the average tax rate among northern counties is 5 points higher than among southern counties, but it is 15 points higher (in northern counties than southern counties) among mountain counties, 5 points higher among valley counties, and

Table 12.12 County Tax Rates by Topography and Section

	Mountain	Valley	Coastal
Southern	5	20	36
	6	19	35
	4	21	34
Northern	20	24	29
	19	25	31
	21	26	30

SOURCE: Hypothetical.

Table 12.13 Mean Tax Rates by Topography and Section

	Mountain	Valley	Coastal	Mean	Effect
Southern	5	20	35	20	−2.5
Northern	20	25	30	25	+2.5
Mean	12.5	22.5	32.5	22.5	
Effect	−10	0	+10		

5 points lower among coastal counties. The effect of section on tax rate now varies with topography.

The interaction sum of squares can be calculated by comparing the actual cell means shown in Table 12.13 with those expected under additivity, as given in Table 12.11. Square the deviation of each cell's mean from that required under additivity, sum those squares, and multiply by the number of cases in each cell. The deviation for the southern mountain cell is $5 - 10 = -5$, that for the northern mountain cell is $20 - 15 = +5$, that for the southern valley cell is $20 - 20 = 0$, for the northern valley cell is $25 - 25 = 0$, for the southern coastal cell is $35 - 30 = 5$, and for the northern coastal cell is $30 - 35 = -5$. The sum of squared deviations is then $(-5)^2 + (5)^2 + 0^2 + 0^2 + (5)^2 + (-5)^2 = 25 + 25 + 0 + 0 + 25 + 25 = 100$. Multiplying this by the three cases in each cell gives an interaction sum of squares of 300. The interaction degrees of freedom is $RC - R - C + 1$, where R is the number of rows (two, here) and C the number of columns (three, here), or $2(3) - 2 - 3 + 1 = 2$. The interaction mean square is the sum of squares for interaction (300) divided by its degrees of freedom (2), or 150.

Note that the data in Table 12.12 happen to have a greater total sum of squares than the earlier tax-rate data analyzed in this chapter, namely 1,624.5. The within sum of squares is the total (1,624.5) minus the columns sum of squares (1,200) minus the rows sum of squares (112.5) minus the interaction sum of squares (300), or 12. The within degrees of freedom is the total (17) minus the column degrees of freedom (2) minus the row degrees of freedom (1) minus the interaction degrees of freedom (2), or 12. The mean square within then is its sum of squares (12) divided by its degrees of freedom (12), or 1.

The within sum of squares and degrees of freedom could also have been computed directly. For the within sum of squares, take every data value in Table 12.12, subtract the mean for that cell as shown in Table 12.13, and sum all the squared deviations to get a total of 12.

Table 12.14 Two-Way Analysis-of-Variance Table: Tax Rate by Topography and Section

Source of Variation	Sum of Squares	*df*	Mean Square (*s*)	*F* Ratio	Required *F* Ratio	Significant?
Columns	1,200	2	600	600	3.88	Yes
Rows	112.5	1	112.5	112.5	4.75	Yes
Interaction	300	2	150	150	3.88	Yes
Within	12	12	1			
Total	1,624.5	17				

For the within degrees of freedom, compute $N - RC$, where N is the number of total observations, R the number of rows, and C the number of columns, or $18 - 2(3) = 12$.

Table 12.14 shows the full two-way analysis-of-variance table. Not only are the column effects and the row effects significant, but the interaction effects are significant. As a result, the only way to describe the data fully would be to indicate the separate effects associated with the columns, the separate effects associated with the rows, and the interaction effects associated with each combination of columns and rows. Occasionally, interaction effects can be significant, even though neither row effects nor column effects are significant, so it is important to test for interaction effects whenever there is more than one case ("replication") for each combination of rows and columns.

SUMMARY

Difference-of-means procedures provide methods for assessing the strength and statistical significance of relationships between nominal independent variables and an interval dependent variable. The choice of procedure depends on the research design—the number of independent variables, the number of categories of each independent variable, how the categories are chosen, whether the independent variables are nested or completely crossed, and whether the design is balanced or unbalanced. The procedures illustrated in this chapter handle the simplest cases; in more complex cases, a statistician should be consulted—preferably at the research design stage, to ensure that the design permits the desired type of analysis.

Questions

1. Say a simple random sample of counties in a state is taken. The tax rates for eastern counties are 10, 20, and 30 while those for the western counties are 40, 50, and 60.
   a. Is the difference of means between west and east statistically significant at the .05 level?
   b. Construct a 95 percent confidence interval.
   c. What proportion of the variance in tax rate is accounted for by region?
   d. Display the results as a one-way analysis of variance.
2. Complete the following analysis-of-variance table for 3 groups and 11 cases per group:

Source of Variation	Sum of Squares	Degrees of Freedom	Mean Square	*F* Ratio	Required *F* Ratio	Significant?
Between	100	______	______	______	______	______
Within	______	______	______			
Total	1,000	______				

13

Research Reports

Research does not end with the data analysis. The final stage is writing up the findings in a report. This should be seen as an integral part of the research process, since the communication of results is important for the sake of scientific progress. Reporting the results permits others to make use of the findings and can suggest the directions that future research should take.

In every field, different authors use different styles of writing, so we do not claim that there is any single correct way to write a research report. However, we shall mention some things which should be kept in mind when writing a report.

SCOPE

A brief report (such as a journal article or a term paper) cannot possibly answer every question that can be asked in the field of study. A report that tries to do so inevitably flounders. Keep the *scope* of your work in mind and remember that a complete treatment of a narrow topic succeeds better than an incomplete treatment of a huge topic. Also, a concrete contribution to the field is more useful than a scattered series of comments.

Choice of Topic

Obviously, the scope of a report is established when the topic is chosen. Suppose we had a choice between a relatively broad and a relatively narrow report. It would be fairly easy to write a report broadly contrasting different groups, such as Protestants and Catholics in terms of their political attitudes. However, such a report would probably find only slight differences and have no way of explaining them. Instead, we might select a single topic of interest (such as the difference in the voting behavior of Protestants and Catholics) and examine it to see how it is affected by a series of controls. This amounts to testing whether the original observed relationship can be explained by a series of alternative hypotheses (for example, to see whether voting differences between respondents of different religions can be explained by differences in their educations, social backgrounds, and so on). The topic is somewhat narrower and is clearly more promising because the controls permit a very detailed analysis of causes and effects.

In choosing a topic, make sure it is one that lends itself to a meaningful statistical analysis. Differences in attitudes toward political protest cannot be studied if there are few differences—there must be enough variance in the dependent variable to make explanation worthwhile. Similarly, there must be differences in the independent variable if it is to explain differences in the dependent variable. Therefore, examine the marginal distributions of the variables (which are often available in a study's codebook) before deciding what can be meaningfully studied. Also, do not limit the group you are studying so severely that statistical controls are impossible. An in-depth analysis of the political attitudes of American Jews is impossible using the data from a single national survey because the number of Jews in a sample would be too small to permit an intensive analysis. (This problem can be solved by combining—sometimes called "pyramiding"—several independent samples to build up enough cases for analysis.) Similarly, many samples do not contain enough blacks to analyze with controls unless special precautions were taken to oversample blacks. Again, check that a meaningful analysis is possible before committing yourself to a topic.

Research Results

The scope of the conclusions must also be kept narrow. In previous chapters, we have stressed how hard it is to prove that a relationship is causal. We have also indicated that the effect of a variable can be gauged properly only if all relevant predictors are included in the model. If all such variables have not been included in the analysis, the interpretations should remain tentative. Additionally, it is important to avoid the temptation to overgeneralize from the results of a data analysis. The results may indicate the nature of a relationship for the data studied, but the relationship might not hold universally. Different results might hold for different countries, or even in different years in the country being studied. Thus, the finding that education affects voting turnout more than sex does (Chapter 8) really means that education had a greater effect on turnout than sex in the United States in 1972; no greater generality for the result can be claimed. Any conclusions reached in a report should be carefully limited to the population and time being studied.

Weak results do not necessarily mean that the topic chosen was poor. The progress of science requires the disproof of expected relationships as much as the discovery of important relationships. Negative findings (findings that show no large relationships) should be reported candidly—there is no need to change topics fast. In any case, do not fall into the statistical fallacy of reporting only the highest correlations. The quality of a research report depends on the coherence of the topic chosen, its theoretical importance, and the imaginativeness of the analysis, not on your luck in finding strong relationships.

ORGANIZATION

Different research reports have different schemes of *organization,* but one scheme is very common. The typical report begins with a statement of the problem to be studied, along with an indication of why the problem is important enough to be studied. This is usually followed by a discussion of the relevant past literature. The author then usually gives his or her own theories as well as the hypotheses to be tested. Sometimes the hypotheses are actually listed and numbered, but a conversational style is more common, with the hypotheses developed as part of the presentation of the problem and theory.

The next major section in most reports is a description of the research method. This means a description of the research operation, including sampling details and sample size. The data analysis follows, with the specific hypotheses tested. Again, there is considerable variation among reports at this stage. Some authors restate their first hypothesis, discuss their test of it, then move to their second hypothesis, and so on. More common is a conversational approach in which the discussion flows logically and hypotheses and results are brought up as they are relevant to the discussion.

After the analysis section usually come sections of "discussion," "conclusions," and/or "summary." These sections review the importance of the results for the original problem, and they indicate what future tests would be appropriate. This last point is important. Rarely does any research effort settle all of the questions on a given topic. The report will summarize the research that has been conducted, after which the author should indicate the total state of research in the area now that this report has been written. The author can indicate what other topics should be studied in the future, what other ways of studying the present topic should be used, how the research design might be modified in the future, and so on. A final summary paragraph might remind the reader of the most important conclusions of the research effort.

STYLE

Finally, it is useful to consider the *style* to be used in technical writing. Research reports usually include large numbers of tables and data displays. Some common sense is useful here—no reader is going to be interested in a tedious report which has sentence after sentence of numbers. The use of statistics does not provide an excuse for an unreadable writing style. If anything, the writing of research reports requires a better writing style than most other writing, since the author must keep the reader's interest while presenting a large amount of numbers. The report should be written in English and in a readable style, without permitting the numbers and data displays to detract from readability.

Writing research reports is actually an art. The author must develop a readable style while finding ways to display data that attract the reader's eye. Well-planned percentage tables are a good idea, and clear figures and charts can be very effective in communicating trends and relationships. Do not put a hundred tables into a research report, no matter how fond you are of them. No reader will survive. (Would you?) Instead of presenting a large number of tables, perhaps one table giving the correlations from each of your separate tables would suffice. Or perhaps a few typical tables might be shown rather than a dozen tables that make the same point.

When writing, try to remember the styles of journal articles that you liked and the styles of articles that you found boring—try to emulate the better-written pieces.

SUMMARY

Research costs a large amount of money, and analysis can take a long time. Given all that expense and effort, it is important to communicate the results in an effective manner. The suggestions given here are just suggestions, and there are other approaches which also work. But it is important to consider questions of scope, organization, and style so that the research report will be an effective communication.

Question

Use one or two pages to write up your analysis of the following table. It is a cross-tabulation of respondents' party identification with whether or not they approve of the president's handling of his job. (Assume the president is a Republican.)

	Partisanship			
President	Republican	Independent	Democrat	Total
Approve	60%	50%	40%	48%
Disapprove	40%	50%	60%	52%
Total	100%	100%	100%	100%
Number of cases	(350)	(420)	(630)	(1,400)

Tau-b = .153 d_{rc} = .135 Gamma = .266 Lambda$_{rc}$ = .104
Tau-c = .174

Answers

Chapter 2

1. The study does not tell what proportion of people who do not disco dance get cancer. It might be that disco dancers get less cancer than would a comparable group of non-disco dancers. A control group is needed.
2. Yale students are not representative of all college students, let alone all young people. For example, they are likely to be much more affluent than the average young person, and more affluent people are more likely to be able to afford travel abroad. The study cannot generalize beyond Yale undergraduates.
3. The study measured tax rates, and that is not a valid indicator of spending on social services. If spending on social services is of interest, then that should be measured more directly.
4. The sampling error for the survey is not reported. There is sampling error associated with any sample, and it is probably at least 3 percent. If it were 3 percent, then the governor's support could be anywhere from 49 percent (52 − 3) to 55 percent (52 + 3). If it were 49 percent, then the governor does not have majority support. Thus unless the sampling error is reported, there is no guarantee from this result that the governor has majority support.

Chapter 3

1.

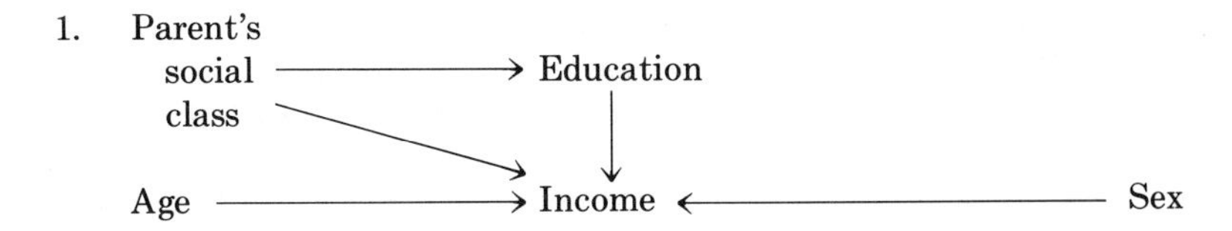

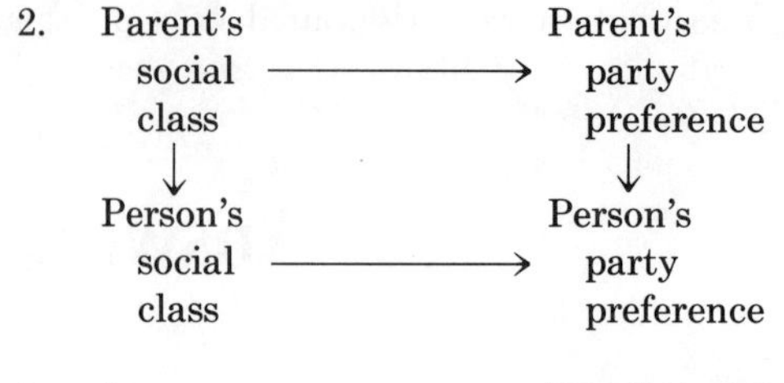

3. Age ⟶ "Traditionalism"
↓
Religion ⟶ Abortion attitude

Chapter 5

1. 937
2. 24.9 percent
3. 42.5 percent
4. 32.7 percent
5. Members of "other" parties, "not ascertaineds," "don't know what parties are."
6. Mode: Democratic; median: Independent, assuming that Independents are "in-between" Republicans and Democrats.
7. $\overline{X} = 80 \qquad s^2 = 100 \qquad s = 10$

Chapter 6

1. Education.
2. Position on war.
3. Yes. The percentages are within categories of the independent variable.
4. c.
5. Hawkish.
6. Small.
7. The number of cases in each educational category.

Chapter 7

1. Race is dichotomous (assuming that races other than white and black are set aside because of too few cases for analysis), and the sign of a correlation between dichotomous variables is meaningless. Therefore, the minus sign does not mean there is a negative relationship between a person's race and interest in politics. It may mean that there is some relationship between race and interest in politics. However, any statement should be qualified, as by saying which race *claims* the greater interest.
2. Religion is a nominal variable (Protestant, Catholic, Jewish, other, none), so a tau correlation should not be calculated. This correlation

would make sense only if the correlation were computed only on the basis of Protestants versus Catholics (or Christians versus Jews, or some religion versus no religion). Otherwise, the lambda coefficient should be used to match the nominal character of religion.

3. Knowledge of the region a person lives in yields a 20 percent reduction in error in guessing the person's vote. A statement of exactly how the regions differ would probably be of more value than this summary statistic.
4. There is a strong tendency for attitudes on abortion and divorce to "scale." Yet gammas can be much larger than taus, so there is no proof of high covariance between these two attitudes.
5. This correlation is very small. Unless the sample is very large, it is not likely to be statistically significant. Even if it is, the correlation is not of major importance. One might believe this correlation if it fits with other known data, but age differences have only a minor effect on attitudes toward divorce according to this result. One possibility to explore is whether age has a steady effect on the attitude. Perhaps the correlation is so small because young people are more tolerant of divorce than middle-aged people, but middle-aged people are less tolerant of divorce than older people. A lambda correlation might test such a possibility.

Chapter 8

1. Reported Turnout in 1972 by Education, Controlling for Sex:

	Males			*Females*		
	Grade School	High School	College	Grade School	High School	College
Voted	67%	74%	85%	50%	68%	89%
Did not vote	33%	26%	15%	50%	32%	11%
Total	100%	100%	100%	100%	100%	100%
N (total 2,281)	(199)	(413)	(362)	(241)	(732)	(334)

SOURCE: 1972 Center for Political Studies American National Election Study

2. The age differences in Table 8.12 are slight, but they are in the direction of younger people being more in favor of an equal role for women than are older people. Table 8.12, however, shows these differences are mainly due to educational differences. There are clear differences between education groups of the same age in their support for an equal role for women. Among the grade-school educated, there is actually a slight tendency for younger people to be more in favor of a traditional role for women than are older people (though this might be due to the small number of young people with only grade-school education, yielding a high sampling error). Among the high-school educated, age has no effect. Only among the college educated are there age differences in the expected direction. There is an education gap but little evidence of a generation gap.

Chapter 9

1. Recode old categories 0, 1, 5, and 6 into new category 2. Recode old categories 2, 3, and 4 into new category 1.
2. New category 1 = 1 on both presidential and congressional vote, or
   = 2 on both presidential and congressional vote.
   New category 2 = 1 on presidential vote and 2 on congressional vote, or
   = 2 on presidential vote and 1 on presidential vote.
   New category 3 = 3 on either presidential or congressional vote.
3. Add up the three items, so that 3 means three campaign activities, 2 means two campaign activities, 1 means one campaign activity, and 0 means no campaign activities. Missing data is a problem here. One solution would be to give a person a score of 9 for missing data if the person has a 9 on any of the three questions (thus, this is not strictly an additive process). Another possibility would be to count the number of campaign activities that the person performed and give the score 9 only if the person had missing data on all three questions.
4. Performed all three activities; attended meetings and did party work, but gave no contributions; attended meetings, but did no party work and gave no contributions; performed none of the three activities. No other patterns fit this scale perfectly.

Chapter 10

1. 30.85 percent.
2. a. $z = (95 - 75/10 = 20/10 = 2$, so .0228.
   b. $75 \pm 1.96\,(10) = 55.4$ to 94.6.
3. a. $10/\sqrt{101 - 1} = 10/\sqrt{100} = 10/10 = 1.0$.
   b. Yes. $z = (78.5 - 75)/1 = 3.5 > 2.58$.
   c. $78.5 \pm 2.58\,(1) = 75.92$ to 81.08.
4. a. $10/\sqrt{26 - 1} = 10/\sqrt{25} = 10/5 = 2.0$.
   b. No. $z = (78.5 - 75)/2 = 1.75 < 1.96$.
   c. $78.5 \pm 1.96\,(2) = 74.58$ to 82.42.
5. a. $10/\sqrt{26 - 1} = 10/\sqrt{25} = 10/5 = 2.0$.
   b. No. $t = (78.5 - 75)/2 = 1.75 < 2.060$, with 25 degrees of freedom.
   c. $78.5 \pm 2.06\,(2) = 74.38$ to 82.62.

Chapter 11

1. $r^2 = 9$ percent.
2. d. Statistical significance does not imply substantive significance.
3. e. It means that a one-unit change in the independent variable induces a one-half-unit change in the dependent variable (when the effects of other variables are *not* controlled).
4. It is impossible to tell from this information. Standardized regression coefficients (beta weights) are required to tell which variable is more important.

Chapter 12

1. a. Yes.

$$s_{\text{diff}} = \sqrt{\frac{100 + 100 + 100 + 100}{4}} \cdot \sqrt{1/3 + 1/3} = \sqrt{100}\sqrt{2/3} = 8.16$$

Critical ratio = 30/8.16 = 3.67, which is greater than $t_{4df} = 2.776$.

b. 30 ± (2.776) (8.16) = 30 ± 22.65 = 7.35 to 52.65.

c. Within sum of squares = 400; sum of squares between = 1350; sum of squares total = 1750. $\eta^2 = 1350/1750 = .77$.

d.

Source of Variation	Sum of Squares	Degrees of Freedom	Mean Square	*F* Ratio	Required *F* at .05	Significant?
Between regions	1,350	1	1,350	13.5	7.71	Yes
Within regions	400	4	100			
Total	1,750	5				

2.

Source of Variation	Sum of Squares	Degrees of Freedom	Mean Square	*F* Ratio	Required *F* at .05	Significant?
Between	100	2	50	1.667	3.32	No
Within	900	30	30			
Total	1,000	32				

Chapter 13

Your answer should point out that, on balance, a majority of Americans do not approve of the president's performance, but that the difference is slight and about the size of sampling error. Furthermore, the difference seems mainly to result from (*a*) greater support for the president among Republicans than Democrats and (*b*) the fact that there are many more Democrats than Republicans.

You should indicate that the effect of partisanship on approval of the president is slight. There is an effect, but it is not very large. You might want to quote the tau value, but you can probably make the point more effectively by indicating that there is a 20 percent difference in support for the president between Republicans and Democrats.

You might also suggest some control variables that would help test this relationship. Region is one of the most likely control variables. For example, support for the president might be well above 40 percent for Southern Democrats though well below 40 percent for Northern Democrats. Thus, partisanship might have a much greater effect on approval of the president for Northerners than for Southerners.

Further Readings

Research Design

Campbell, Donald T., and Stanley, Julian C. *Experimental and Quasi-Experimental Designs for Research.* Chicago: Rand-McNally, 1963.

Kerlinger, Fred. *Foundations of Behavioral Research.* 2nd ed. New York, Holt, Rinehart, & Winston, 1973.

Kish, Leslie. "Some Statistical Problems on Research Design," *American Sociological Review, 24* (1959), pp. 328–38.

McGaw, Dickinson L., and Watson, George L. *Political and Social Inquiry.* New York: Wiley, 1976.

Selltiz, Claire, Wrightsman, Lawrence S., and Cook, Stuart W. *Research Methods in Social Relations.* 3rd ed. New York: Holt, Rinehart, & Winston, 1976.

Webb, Eugene J., Campbell, Donald T., Schwartz, Richard D., and Sechrest, Lee. *Unobtrusive Measures: Nonreactive Research in the Social Sciences.* Chicago: Rand-McNally, 1966.

Sampling

Kish, Leslie. *Survey Sampling.* New York: Wiley, 1965.

Moser, C. A., and Kalton, Graham, *Survey Methods in Social Investigations.* 2nd ed. New York: Basic Books, 1972.

Sudman, Seymour. *Applied Sampling.* New York: *Academic Press,* 1976.

Computer Utilization

Janda, Kenneth. *Data Processing: Applications to Political Research.* 2nd ed. Evanston: Northwestern University Press, 1969.

Nie, Norman H., Hull, C. Hadlai, Jenkins, Jean K., Steinbrenner, Karin, and Bent, Dale H. *SPSS: Statistical Package for the Social Sciences.* 2nd ed. New York: McGraw-Hill, 1975.

Basic Statistics

Blalock, Hubert, M., Jr. *Social Statistics.* 2nd ed. New York: McGraw-Hill, 1972.

Hanushek, Eric A., and Jackson, John E. *Statistical Methods for Social Scientists.* New York: Academic Press, 1977.

Hays, William L. *Statistics for the Social Sciences.* 2nd ed. New York: Holt, Rinehart, & Winston, 1973.

Probability

Feller, William. *An Introduction to Probability Theory and Its Applications.* Vol. 1, 2nd ed. New York: Wiley, 1957.

Parzen, Emanuel. *Modern Probability Theory and Its Applications.* New York: Wiley. 1960.

Statistical Inference

Morrison, Denton E., and Henkel, Ramon E. (Eds.). *The Significance Test Controversy.* Chicago: Aldine-Atherton, 1970.

Measures of Association

Costner, Herbert L. "Criteria for Measures of Association," *American Sociological Review, 30* (1965), pp. 341–53.

Freeman, Linton C. *Elementary Applied Statistics for Students in Behavioral Science.* New York: Wiley, 1965.

Reynolds, H. T. *The Analysis of Cross-Classifications.* New York: Free Press, 1977.

Siegel, Sidney. *Nonparametric Statistics for the Behavioral Sciences.* New York: McGraw-Hill, 1956.

Weisberg, Herbert F. "Models of Statistical Relationship," *American Political Science Review, 68* (1974), pp. 1638–55.

Weiss, Robert. *Statistics for Social Research.* New York: Wiley, 1968.

Statistical Controls

Rosenberg, Morris. *Logic of Survey Analysis.* New York: Basic Books, 1968.

Scaling

Maranell, Gary M. (Ed.) *Scaling: A Sourcebook for Behavioral Scientists.* Chicago: Aldine, 1974.

Rabinowitz, George B. "An Introduction to Nonmetric Multidimensional Scaling," *American Journal of Political Science, 19* (1975), pp. 343–90.

Weisberg, Herbert G., and Rusk, Jerrold G. "Dimensions of Candidate Evaluation," *American Political Science Review, 64* (1970), pp. 1167–85.

Regression Analysis

Draper, Norman R., and Smith, Harry. *Applied Regression Analysis.* New York: Wiley, 1966.
Edwards, Allen L. *An Introduction to Linear Regression and Correlation.* San Francisco: W. H. Freeman and Company, 1976.
Tufte, Edward R. *Data Analysis for Politics and Policy.* Englewood Cliffs, N.J.: Prentice-Hall, 1974.

Causal Modeling

Asher, Herbert B. *Causal Modeling.* Beverly Hills, Calif.: Sage, 1976.
Blalock, Hubert M., Jr. *Causal Inferences in Nonexperimental Research.* Chapel Hill: University of North Carolina Press, 1964.
Blalock, Hubert M., Jr. (Ed.). *Causal Models in the Social Sciences.* Chicago: Aldine-Atherton, 1971.

Factor Analysis

Harman, Harry H. *Modern Factor Analysis.* 2nd ed. Chicago: University of Chicago Press, 1967.
Kim, Jae-On, and Charles W. Mueller. "Introduction to Factor Analysis." Beverly Hills, Calif.: Sage, 1978.

Analysis of Variance

Iversen, Gudmund R., and Norpoth, Helmut. *Analysis of Variance.* Beverly Hills, Calif.: Sage, 1976.
Lindman, Harold R. *Analysis of Variance in Complex Experimental Designs.* San Francisco: W. H. Freeman and Company, 1974.
Scheffé, Henry. *The Analysis of Variance.* New York: Wiley, 1959.

Index to Notation and Statistics: Terms and Symbols

Level of Measurement

Nominal (unordered categories)	
Central tendency	Mode, 52
Measure of association	λ (lambda)—predictive, 82
Significance test	χ^2 (chi square), 86
Ordinal (ordered categories)	
Central tendency	Median, 51
Measures of association	*d*—predictive, 77
	γ (gamma)—scale, 78
	r_s (rho)—rank orders, 81
	τ (tau)—covariation, 75
Interval (numeric)	
Central tendency	$\overline{X}$, Mean, 50
Dispersion	s, Standard deviation, 55
	s_{diff}, Standard error of the difference, 171
	s_{m}, Standard error of mean, 56
	s^2, Variance, 53
	s^2_Y, Residual variance, 148
Measures of relationship	r, Correlation, 143
	a, Regression intercept, 146
	b, Regression slope, 146
	β (beta), Standardized regression coefficient, 159
	η^2 (eta-squared), Correlation ratio, 175
	R, Multiple correlation coefficient, 158
Significance tests	Correlation, 151
	Difference in proportions, 63
	Mean, 55
	Multiple correlation coefficient, 160
	Proportion, 14
	Regression coefficient, 159
	Difference-of-means test, 170
	One-way analysis of variance, 179
	Two-way analysis of variance, 185

Other Statistics

Coefficient of reproducibility, 108
Coefficient of scalability, 108
Communality, 166
Minimum marginal reproducibility, 108
Sampling error, 14
z scores, 118
p value, 132
Confidence interval, 132
t value, 136
F ratio, 180

Other Notation

df Degrees of freedom, 88
N Number of cases, 39
% Percentage, 39
Σ (sigma) Summation, 51
k Number of independent variable categories, 181
R Number of categories of row variable, 193
C Number of categories of column variable, 193

Index

Affirming the consequent, fallacy of, 126
Analysis design, 25–27
Analysis of covariance, 185
Analysis of variance, 179; one-way, 179–185; two-way, 185–194; table, 181–183; *F* test of, 183; assumptions, 185; types of models, 186; interaction, 187, 192–194; additivity, 190–192
Association measures, 69; alternative, 74–75; choice of, 79–80; 84–85; and dependent variable, 77–78, 82; features of, 72–74; and levels of measurement, 75, 80–81, 141–142; logic of, 69–75; models of, 69–70; nominal, 82–85; ordinal, 75–82; sign of, 70, 72–74, 84; size of, 63, 66–67, 74, 76. *See also* Percentages, differences
Attitudes: differences in, 66–67, 197; in explanations, 20
Average. *See* Central tendency

Behaviors: in explanations, 20
Blalock, Hubert M., 19n
Brown, Thad A., 163n
Bunge, Mario, 19n

Causal modeling, 23, 162–163
Causal process, 22–24, 157
Cause, 19; conditions for, 20–22; and explanation, 20
Central-limit theorem, 123–126
Central tendency, 49; choice of measure, 52; measures of, 49–53
Coding, 29; effects on measures of association, 73–74
Communality, 166
Computer: advantages of, 28–29, 31; instructions, 33–36; limitations of, 32–33, 52; modes of operation, 33; programs, 32–36; required information, 35–36; use, 28–37
Computer card, 29–31; problems with, 30–31
Computer tapes, 31
Confidence intervals, 46, 132–135; for difference of means, 172–173
Controls, 7; choice of, 100–101; for correlations, 154–157; direction, 98–99; interpretation, 94–98; multiple, 100–101; statistical, 90–102, 156–157; for tables, 90–102; and theory, 90, 100–101
Correlated predictors, 157, 160
Correlation: controls for, 154–157; meaning of, 19, 89; and theory, 89
Correlation coefficient, 69, 143; calculation of, 147–148; interpretation of, 147–150; and levels of measurement, 80–81, 141–142; and multidimensional scaling, 112–114; multiple, 158; partial, 155–158
Costner, Herbert T., 79n
Covariation, 69–70, 75–77
Critical ratio, 130; for difference of means, 170–172
Critical region, 130–131
Cross-tabulations, 58–60
Cumulation, 109–111
Curvilinear relationship, 80–81, 149–151, 160

Data: archives, 25; missing, 40–43, 52–53, 64, 80, 107, 110–111; storage, 28–29, 30–31
Degrees of freedom, 86–88; *t*-distribution, 136; for analysis of variance, 180–181; for *F* test, 183

Demographic variables: and explanation, 20
Dependent variable, 19–20, 60–62; and measures of association, 77–78, 82
Dichotomous variables, 49, 80, 84, 160–161
Dimensions: naming, 114; number of, 107–108, 111–112, 164, 166. *See also* Factor analysis; Guttman scales; Multidimensional scaling; Scale relationship
Directional tests, 134–135
Don't knows, 41–43
Dummy variables, 160–161

Errors: in describing a relationship, 59–60; in Guttman scales, 109–111; proportional reduction in 83, 147, 175; in rounding percentages, 41; types of, 128–129; Type I and Type II, 128–129
Eta-squared, 175–179; calculation of, 176–177; in calculation of F, 183
Exhaustive, 117
Experiments, 8–9; alternatives to, 92
Explanation, 18–22

F or *F* test, 159, 183; table of *F* distribution, 182
Factor analysis, 163–167; interpretation, 163–164
Factor loadings, 166
Frequency distribution, 38–46; interpretation, 43–46

Gallup, George, 25n
Goodman, Leo, 78
Graphs, 65–66, 99, 150
Grether, David M., 142n
Guttman scales, 108–112

Henkel, Ramon E., 88n
Hypothesis, 18–19, 26, 200–201; alternative, 21–22, 26, 199
Hypothesis testing, 126–132; null, 127; research, 127

Independent variables, 20, 60–62; relative importance of, 159
Indicies, 106–108; dimensionality of, 107–108
Interaction, 99–100, 160
Intercept, 143, 146, 158
Inter-University Consortium for Political and Social Research, 25
Interval data, 141–142
Interval statistics, 140–166; limitations of, 167

Jackson, John E., 23n

Kendall, Maurice, 75
Keypunching, 30
Kruskal, J. B., 114
Kruskal, W. H., 78

Labovitz, Sanford, 142n
Law of large numbers, 124
Least-squares, 145–146, 158
Levels of measurement, 46–49; and central tendency measures, 51–52; choice of, 49; and dispersion, 55; measures of association, 75, 80–81, 141–142
Linear relationship, 142–144, 145–147, 149, 160
Lininger, Charles A., 63
Logic of analysis, 3

Marginals. *See* Frequency distribution
Mean(s), 49–52; comparing, 168–173; confidence interval for, 172–173; critical ratio for, 170–172; standard error of, 56–57, 171
Mean squares, 180
Measurement, 6–7
Measures of association. *See* Association measures
Miller, Arthur H., 163n
Miller, Warren E., 163n
Missing data. *See* Data, missing
Morrison, Denton E., 88n
Multicollinearity, 160
Multidimensional scaling, 112–114. *See also* Factor analysis
Multiple correlation coefficient, 158
Multiple regression, 157–163; interpretation of, 159–162; procedure, 157–158
Mutually exclusive, 117

Nonattitudes, 41–43
Normal distribution, 55–56, 117–118; areas under, 120–121; assumption, 151, 159; importance of, 121–123; table of, 122

Partial correlation coefficient, 155–157
Pearson, Karl, 143
Percentages: comparing, 60–66; differences, 60–66, 77–78; directions, 61–62; distributions, 39; tables, 59–60

Planning research, 4
Predictability model of a relationship, 70, 81, 82–84, 149, 175
Presidential vote, 29, 44–46, 64–66, 83–84, 105–106, 162–163
Probability, 116–117; sampling, 13; value, 131–132
Proportional reduction in error (PRE), 83, 147, 175

Raine, Alden S., 163n
Randomization: definition, 7–8
Rank orders, 81–82
Reciprocal causation, 21, 23–24
Recoding, 103–106; bivariate, 105–106
Regression, 142–154; interpretation, 149–154; multiple, 157–163; procedure, 144–148
Regression coefficient: calculation, 146; definition, 143; partial, 158; standardized, 159
Relationship: interpretation of, 58–68; models of, 69–70, 81, 82–84, 149; scale, 78, 108–111; sign of, 69–70; size of, 62–63, 66–68, 69, 70–72; specifying of, 94, 95–96, 156–157. *See also* Association measures; Correlation
Reliability, 11
Representation, 8
Research: interpretation of, 2; evaluation of, 3
Research design, 6–17
Research report, 196–200
Residuals, 148, 150, 155, 157–158
Residual variance, 145–148, 158, 176
Rusk, Jerrold G., 113

Sample, 9–10; size, 15, 56; types of, 13–14
Sampling distribution, 123
Sampling error, 14–15, 43–44, 55–57; computation, 55–57; and relationship, 66
Sampling fraction, 15
Sampling procedures, 12–16
Scale relationship, 78, 108–111
Scatter diagram, 150
Scientific method, 18–24
Secondary analysis, 24–25, 26
Sign of relationship, 69–70
Significance tests, 85–89, 173; and chance, 153–154; and populations, 32; problems with, 32, 56–57, 88, 152–154; and sample size, 32, 62–63, 88, 152–153; and sampling method, 56–57, 88, 151; and substantive significance, 88, 153
Simple random sample, 13
Slope. *See* Regression coefficient
Somers, Robert, 77
Spearman, Charles, 81
Specifying a relationship, 94, 97–98, 156–157
Spurious relationships, 91–94, 95–97, 156–157
Standard error of the mean, 56–57, 123; of difference of means, 171; of regression coefficient, 159
Statistical control, 7
Statistical independence, 71
Statistical Package for the Social Sciences (SPSS), 35, 36, 104, 110
Statistics, 5
Strength of relationship, 174–179

t-distribution, 125, 136–138; table of, 137
t-tests, 159, 173, 184
Tables: controls for, 90–102; cross-tabulations, 58–60
Terminals, 33
Theory, role of, 23–24, 89
Third variables. *See* Controls

Unbiasedness, 12
Unidimensionality. *See* Dimensions
Unobtrusive data, 10–11

Validity, 11–12
Variables, 19–20, 25–26; dependent, 19–20, 60–62, 77–78, 82; dichotomous, 49, 80, 84, 160; dummy, 160; independent, 20, 60–62, 159; need for variance in, 55, 197
Variance, 53–55, 145; explained, 143, 147–148, 158; residual, 145–147, 158, 175
Vectors, 163–164, 166
Verifying, 30
Voting turnout, 45–46, 61–63, 90, 95–101

Warwick, Donald P., 63
Weisberg, Herbert F., 79n, 113

z scores, 118